Verification & Validation of Selected Fire Models for Nuclear Power Plant Applications

Volume 1: Main Report

NUREG-1824   **EPRI 1011999**

Final Report

May 2007

U.S. Nuclear Regulatory Commission
Office of Nuclear Regulatory Research (RES)
Two White Flint North, 11545 Rockville Pike
Rockville, MD 20852-2738

U.S. NRC-RES Project Manager
M. H. Salley

Electric Power Research Institute (EPRI)
3420 Hillview Avenue
Palo Alto, CA 94303

EPRI Project Manager
R.P. Kassawara

DISCLAIMER OF WARRANTIES AND LIMITATION OF LIABILITIES

NOTE

CITATIONS

This report was prepared by:

U.S. Nuclear Regulatory Commission,
Office of Nuclear Regulatory Research (RES)
Two White Flint North, 11545 Rockville Pike
Rockville, MD 20852-2738
Principal Investigators:
K. Hill
J. Dreisbach

Electric Power Research Institute (EPRI)
3420 Hillview Avenue
Palo Alto, CA 94303
Science Applications International Corp (SAIC)
4920 El Camino Real
Los Altos, CA 94022
Principal Investigators:
F. Joglar
B. Najafi

National Institute of Standards and Technology
Building Fire Research Laboratory (BFRL)
100 Bureau Drive, Stop 8600
Gaithersburg, MD 20899-8600

Principal Investigators:

K McGrattan

R. Peacock

A. Hamins

Volume 1, Main Report: B. Najafi, F. Joglar, J. Dreisbach
Volume 2, Experimental Uncertainty: A. Hamins, K. McGrattan
Volume 3, FDT^S: J. Dreisbach, K. Hill
Volume 4, FIVE-Rev1: F. Joglar
Volume 5, CFAST: R. Peacock, P. Reneke (NIST)
Volume 6, MAGIC: F. Joglar, B. Guatier (EdF), L. Gay (EdF), J. Texeraud (EdF)
Volume 7, FDS: K. McGrattan

This report describes research sponsored jointly by U.S. Nuclear Regulatory Commission, Office of Nuclear Regulatory Research (RES) and Electric Power Research Institute (EPRI).

The report is a corporate document that should be cited in the literature in the following manner:

Verification and Validation of Selected Fire Models for Nuclear Power Plant Applications, Volume 1: Main Report, U.S. Nuclear Regulatory Commission, Office of Nuclear Regulatory Research (RES), Rockville, MD, 2007, and Electric Power Research Institute (EPRI), Palo Alto, CA, NUREG-1824 and EPRI 1011999.

ABSTRACT

There is a movement to introduce risk-informed and performance-based analyses into fire protection engineering practice, both domestically and worldwide. This movement exists in the general fire protection community, as well as the nuclear power plant (NPP) fire protection community. The U.S. Nuclear Regulatory Commission (NRC) has used risk-informed insights as part of its regulatory decision making since the 1990's.

In 2002, the National Fire Protection Association (NFPA) developed NFPA 805, *Performance-Based Standard for Fire Protection for Light-Water Reactor Electric Generating Plants, 2001 Edition*. In July 2004, the NRC amended its fire protection requirements in Title 10, Section 50.48, of the *Code of Federal Regulations* (10 CFR 50.48) to permit existing reactor licensees to voluntarily adopt fire protection requirements contained in NFPA 805 as an alternative to the existing deterministic fire protection requirements. In addition, the NPP fire protection community has been using risk-informed, performance-based (RI/PB) approaches and insights to support fire protection decision-making in general.

One key tool needed to further the use of RI/PB fire protection is the availability of verified and validated fire models that can reliably predict the consequences of fires. Section 2.4.1.2 of NFPA 805 requires that only fire models acceptable to the Authority Having Jurisdiction (AHJ) shall be used in fire modeling calculations. Furthermore, Sections 2.4.1.2.2 and 2.4.1.2.3 of NFPA 805 state that fire models shall only be applied within the limitations of the given model, and shall be verified and validated.

This report is the first effort to document the verification and validation (V&V) of five fire models that are commonly used in NPP applications. The project was performed in accordance with the guidelines that the American Society for Testing and Materials (ASTM) set forth in ASTM E 1355, *Standard Guide for Evaluating the Predictive Capability of Deterministic Fire Models*. The results of this V&V are reported in the form of ranges of accuracies for the fire model predictions.

FOREWORD

Fire modeling and fire dynamics calculations are used in a number of fire hazards analysis (FHA) studies and documents, including fire risk analysis (FRA) calculations; compliance with and exemptions to the regulatory requirements for fire protection in 10 CFR Part 50; the Significance Determination Process (SDP) used in the inspection program conducted by the U.S. Nuclear Regulatory Commission (NRC); and, most recently, the risk-informed performance-based (RI/PB) voluntary fire protection licensing basis established under 10 CFR 50.48(c). The RI/PB method is based on the National Fire Protection Association (NFPA) Standard 805, *Performance-Based Standard for Fire Protection for Light-Water Reactor Generating Plants*.

The seven volumes of this NUREG-series report provide technical documentation concerning the predictive capabilities of a specific set of fire dynamics calculation tools and fire models for the analysis of fire hazards in postulated nuclear power plant (NPP) scenarios. Under a joint memorandum of understanding (MOU), the NRC Office of Nuclear Regulatory Research (RES) and the Electric Power Research Institute (EPRI) agreed to develop this technical document for NPP application of these fire modeling tools. The objectives of this agreement include creating a library of typical NPP fire scenarios and providing information on the ability of specific fire models to predict the consequences of those typical NPP fire scenarios. To meet these objectives, RES and EPRI initiated this collaborative project to provide an evaluation, in the form of verification and validation (V&V), for a set of five commonly available fire modeling tools.

The road map for this project was derived from NFPA 805 and the American Society for Testing and Materials (ASTM) Standard E 1355, *Standard Guide for Evaluating the Predictive Capability of Deterministic Fire Models*. These industry standards form the methodology and process used to perform this study. Technical review of fire models is also necessary to ensure that those using the models can accurately assess the adequacy of the scientific and technical bases for the models, select models that are appropriate for a desired use, and understand the levels of confidence that can be attributed to the results predicted by the models. This work was performed using state-of-the-art fire dynamics calculation methods/models and the most applicable fire test data. Future improvements in the fire dynamics calculation methods/models and additional fire test data may impact the results presented in the seven volumes of this report.

This document does not constitute regulatory requirements, and NRC participation in this study neither constitutes nor implies regulatory approval of applications based on the analysis contained in this text. The analyses documented in this report represent the combined efforts of individuals from RES and EPRI. Both organizations provided specialists in the use of fire models and other FHA tools to support this work. The results from this combined effort do not constitute either a regulatory position or regulatory guidance. Rather, these results are intended to provide technical analysis of the predictive capabilities of five fire dynamic calculation tools, and they may also help to identify areas where further research and analysis are needed.

Brian W. Sheron, Director
Office of Nuclear Regulatory Research
U.S. Nuclear Regulatory Commission

CONTENTS

FIGURES

TABLES

REPORT SUMMARY

This report documents the verification and validation (V&V) of five selected fire models commonly used in support of risk-informed and performance-based (RI/PB) fire protection at nuclear power plants (NPPs).

Background

Since the 1990s, when it became the policy of the NRC to use risk-informed methods to make regulatory decisions where possible, the nuclear power industry has been moving from prescriptive rules and practices toward the use of risk information to supplement decision-making. Several initiatives have furthered this transition in the area of fire protection. In 2001, the National Fire Protection Association (NFPA) completed the development of NFPA Standard 805, *Performance-Based Standard for Fire Protection for Light-Water Reactor Electric Generating Plants*, 2001 Edition. Effective July 16, 2004, the NRC amended its fire protection requirements in Title 10, Section 50.48(c), of the *Code of Federal Regulations* [10 CFR 50.48(c)] to permit existing reactor licensees to voluntarily adopt fire protection requirements contained in NFPA 805 as an alternative to the existing deterministic fire protection requirements. RI/PB fire protection often relies on fire modeling for determining the consequence of fires. NFPA 805 requires that the "fire models shall be verified and validated," and "only fire models that are acceptable to the Authority Having Jurisdiction (AHJ) shall be used in fire modeling calculations."

Objectives

- To perform V&V studies of selected fire models using a consistent methodology (ASTM I 1335)

- To investigate the specific fire modeling issue of interest to NPP fire protection applications

- To quantify fire model predictive capabilities to the extent that can be supported by comparison with selected and available experimental data.

Approach

This project team performed V&V studies on five selected models: (1) NRC's NUREG-1805 Fire Dynamics Tools (FDTS), (2) EPRI's Fire-Induced Vulnerability Evaluation Revision 1 (FIVE-Rev1), (3) National Institute of Standards and Technology's (NIST) Consolidated Model of Fire Growth and Smoke Transport (CFAST), (4) Electricité de France's (EdF) MAGIC, and (5) NIST's Fire Dynamics Simulator (FDS). The team based these studies on the guidelines of the ASTM E 1355, *Standard Guide for Evaluating the Predictive Capability of Deterministic Fire Models*. The scope of these V&V studies was limited to the capabilities of the selected fire models and did not cover certain potential fire scenarios that fall outside the capabilities of these fire models.

Results

The results of this study are presented in the form of relative differences between fire model predictions and experimental data for fire modeling attributes such as plume temperature that are important to NPP fire modeling applications. While the relative differences sometimes show agreement, they also show both under-prediction and over-prediction in some circumstances. These relative differences are affected by the capabilities of the models, the availability of accurate applicable experimental data, and the experimental uncertainty of these data. The project team used the relative differences, in combination with some engineering judgment as to the appropriateness of the model and the agreement between model and experiment, to produce a graded characterization of each fire model's capability to predict attributes important to NPP fire modeling applications.

This report does not provide relative differences for all known fire scenarios in NPP applications. This incompleteness is attributable to a combination of model capability and lack of relevant experimental data. The first problem can be addressed by improving the fire models, while the second problem calls for more applicable fire experiments.

EPRI Perspective

The use of fire models to support fire protection decision-making requires a good understanding of their limitations and predictive capabilities. While this report makes considerable progress toward this goal, it also points to ranges of accuracies in the predictive capability of these fire models that could limit their use in fire modeling applications. Use of these fire models presents challenges that should be addressed if the fire protection community is to realize the full benefit of fire modeling and performance-based fire protection. Persisting problems require both short-term and long-term solutions. In the short-term, users need to be educated on how the results of this work may affect known applications of fire modeling, perhaps through pilot application of the findings of this report and documentation of the resulting lessons learned. In the long-term, additional work on improving the models and performing additional experiments should be considered.

Keywords

Fire
Verification and Validation (V&V)
Risk-Informed Regulation
Fire Safety
Nuclear Power Plant
Fire Probabilistic Safety Assessment (PSA)

Fire Modeling
Performance-Based
Fire Hazard Analysis (FHA)
Fire Protection
Fire Probabilistic Risk Assessment (PRA)

PREFACE

This report is presented in seven volumes. Volume 1, the Main Report, provides general background information, programmatic and technical overviews, and project insights and conclusions. Volume 2 quantifies the uncertainty of the experiments used in the V&V study of the five fire models considered in this study. Volumes 3 through 7 provide detailed discussions of the verification and validation (V&V) of the following fire models:

Volume 3   Fire Dynamics Tools (FDT^s)

Volume 4   Fire-Induced Vulnerability Evaluation, Revision 1 (FIVE-Rev1)

Volume 5   Consolidated Model of Fire Growth and Smoke Transport (CFAST)

Volume 6   MAGIC

Volume 7   Fire Dynamics Simulator (FDS)

ACKNOWLEDGMENTS

The work documented in this report benefited from contributions and considerable technical support from several organizations.

The verification and validation (V&V) studies for FDT^s (Volume 3), CFAST (Volume 5), and FDS (Volume 7) were conducted in collaboration with the U.S. Department of Commerce, National Institute of Standards and Technology (NIST), Building and Fire Research Laboratory (BFRL). Since the inception of this project in 1999, the NRC has collaborated with NIST through an interagency memorandum of understanding (MOU) and conducted research to provide the necessary technical data and tools to support the use of fire models in nuclear power plant fire hazard analysis (FHA).

We appreciate the efforts of Doug Carpenter and Rob Schmidt of Combustion Science Engineers, Inc. for their comments and contributions to Volume 3.

In addition, we acknowledge and appreciate the extensive contributions of Electricité de France (EdF) in preparing Volume 6 for MAGIC.

We thank Drs. Charles Hagwood and Matthew Bundy of NIST for the many helpful discussions regarding Volume 2.

We also appreciate the efforts of organizations participating in the International Collaborative Fire Model Project (ICFMP) to Evaluate Fire Models for Nuclear Power Plant Applications, which provided experimental data, problem specifications, and insights and peer comment for the international fire model benchmarking and validation exercises, and jointly prepared the panel reports used and referred to in this study. We specifically appreciate the efforts of the Building Research Establishment (BRE) and the Nuclear Installations Inspectorate in the United Kingdom, which provided leadership for ICFMP Benchmark Exercise (BE) #2, as well as Gesellschaft für Anlagen-und Reaktorsicherheit (GRS) and Institut für Baustoffe, Massivbau und Brandschutz (iBMB) in Germany, which provided leadership and valuable experimental data for ICFMP BE #4 and BE #5. In particular, ICFMP BE #2 was led by Stewart Miles at BRE; ICFMP BE #4 was led by Walter Klein-Hessling and Marina Rowekamp at GRS, and R. Dobbernack and Olaf Riese at iBMB; and ICFMP BE #5 was led by Olaf Riese and D. Hosser at iBMB, and Marina Rowekamp at GRS. Simo Hostikka of VTT, Finland also assisted with ICFMP BE#2 by providing pictures, tests reports, and answered various technical questions of those experiments. We acknowledge and sincerely appreciate all of their efforts.

We greatly appreciate Paula Garrity, Technical Editor for the Office of Nuclear Regulatory Research, and Linda Stevenson, agency Publications Specialist, for providing editorial and publishing support for this report. Lionel Watkins and Felix Gonzalez developed the graphics for Volume 1. We also greatly appreciate Dariusz Szwarc and Alan Kouchinsky for their assistance finalizing this report.

We wish to acknowledge the team of peer reviewers who reviewed the initial draft of this report and provided valuable comments. The peer reviewers were Dr. Craig Beyler and Mr. Phil DiNenno of Hughes Associates, Inc., and Dr. James Quintiere of the University of Maryland.

Finally, we would like to thank the internal and external stakeholders who took the time to provide comments and suggestions on the initial draft of this report when it was published in the *Federal Register* (71 FR 5088) on January 31, 2006. Those stakeholders who commented are listed and acknowledged below.

Janice Bardi, ASTM International

Moonhak Jee, Korea Electric Power Research Institute

U.S. Nuclear Regulatory Commission, Office of Nuclear Reactor Regulation Fire Protection Branch

J. Greg Sanchez, New York City Transit

David Showalter, Fluent, Inc.

Douglas Carpenter, Combustion Science & Engineering, Inc.

Nathan Siu, U.S. Nuclear Regulatory Commission, Office of Nuclear Regulatory Research

Clarence Worrell, Pacific Gas & Electric

LIST OF ACRONYMS

AGA	American Gas Association
AHJ	Authority Having Jurisdiction
ASME	American Society of Mechanical Engineers
ASTM	American Society for Testing and Materials
BE	Benchmark Exercise
BFRL	Building and Fire Research Laboratory
BRE	Building Research Establishment
BWR	Boiling-Water Reactor
CDF	Core Damage Frequency
CFAST	Consolidated Fire Growth and Smoke Transport Model
CFD	Computational Fluid Dynamics
CFR	*Code of Federal Regulations*
CSR	Cable Spreading Room
EdF	Electricité de France
EPRI	Electric Power Research Institute
FDS	Fire Dynamics Simulator
FDT^s	Fire Dynamics Tools (NUREG-1805)
FHA	Fire Hazard Analysis
FIVE-Rev1	Fire-Induced Vulnerability Evaluation, Revision 1
FM/SNL	Factory Mutual & Sandia National Laboratories
FPA	Foote, Pagni, and Alvares
FRA	Fire Risk Analysis
GRS	Gesellschaft fuer Anlagen-und Reaktorsicherheit (Germany)
HGL	Hot Gas Layer
HRR	Heat Release Rate

IAFSS	International Association of Fire Safety Science
iBMB	Institut für Baustoffe, Massivbau und Brandschutz
ICFMP	International Collaborative Fire Model Project
IEEE	Institute of Electrical and Electronics Engineers
IPEEE	Individual Plant Examination of External Events
MCC	Motor Control Center
MCR	Main Control Room
MQH	McCaffrey, Quintiere, and Harkleroad
MOU	Memorandum of Understanding
NBS	National Bureau of Standards (now NIST)
NFPA	National Fire Protection Association
NIST	National Institute of Standards and Technology
NPP	Nuclear Power Plant
NRC	U.S. Nuclear Regulatory Commission
NRR	Office of Nuclear Reactor Regulation (NRC)
PMMA	Polymethyl-methacrylate
PRA	Probabilistic Risk Assessment
PWR	Pressurized-Water Reactor
RCP	Reactor Coolant Pump
RES	Office of Nuclear Regulatory Research (NRC)
RI/PB	Risk-Informed, Performance-Based
SBDG	Stand-By Diesel Generator
SDP	Significance Determination Process
SFPE	Society of Fire Protection Engineers
SNL	Sandia National Laboratory
SWGR	Switchgear Room
V&V	Verification & Validation

1
INTRODUCTION

1.1 Background

Over the past decade, there has been a considerable movement in the nuclear power industry to transition from prescriptive rules and practices toward the use of risk information to supplement decision-making. In the area of fire protection, this movement is evidenced by numerous initiatives by the U.S. Nuclear Regulatory Commission (NRC) and the nuclear power generation community worldwide.

In 2001, the National Fire Protection Association (NFPA) completed its development of NFPA 805, *Performance-Based Standard for Fire Protection for Light-Water Reactor Electric Generating Plants*, 2001 Edition [1]. Effective July 16, 2004, the NRC amended its fire protection requirements in Title 10, Section 50.48(c), of the *Code of Federal Regulations* [10 CFR 50.48(c)] to permit existing reactor licensees to voluntarily adopt fire protection requirements contained in NFPA 805 as an alternative to the existing deterministic fire protection requirements [2].

Risk-informed, performance-based (RI/PB) fire protection often relies on fire modeling to determine the consequences of fires. NFPA 805 states that "fire models shall be verified and validated," and "only fire models that are acceptable to the authority having jurisdiction (AHJ) shall be used in fire modeling calculations."

1.2 Programmatic Overview

Under a Memorandum of Understanding [3], the NRC's Office of Nuclear Regulatory Research (RES) and the Electric Power Research Institute (EPRI) initiated a collaborative project for verification and validation (V&V) of five selected fire models to support RI/PB fire protection and implementation of the voluntary fire protection rule that adopts NFPA 805 as an RI/PB alternative. This V&V effort may also serve to increase the confidence of reviewers who evaluate fire models that are used in other programs, such as the Fire Protection Significance Determination Process (SDP).

This collaboration brings together the combined information and knowledge generated in this area by the NRC and EPRI fire research programs. The National Institute of Standards and Technology (NIST) was also an important partner in this project. NIST provided extensive modeling and experimentation expertise. This effort also recognizes the considerable knowledge that resides in the fire science community in general, and attempts to use that knowledge, particularly within the context of the fire models being evaluated. This report is the direct result of this collaboration between RES, EPRI, and NIST.

1.2.1 Objectives

The purpose of this report is to describe an evaluation of the predictive capabilities of certain fire models for applications specific to nuclear power plants (NPPs). These models may be used to demonstrate compliance with the requirements of 10 CFR 50.48(c) [2] and the referenced NFPA standard, NFPA 805 [1].

Engineering analyses and methods that are applied to demonstrate compliance with the performance criteria in NFPA 805 need the requisite degree of defensible technical justification, as dictated by the scope and complexity of the specific application. These analyses should be performed by qualified analysts and should include any necessary V&V of analytical methods relevant to the specific application.

Section 2.4.1.2 of NFPA 805 states that only fire models acceptable to the AHJ shall be used in fire modeling calculations. Further, Sections 2.4.1.2.2 and 2.4.1.2.3 of NFPA 805 state that fire models shall only be applied within the limitations of the given fire model, and shall be verified and validated. Thus, V&V is necessary to establish acceptable uses and limitations of fire models. In addition, analysts need to justify the appropriateness of fire model for specific applications.

Verification and validation of a calculation method are intended to ensure the correctness and suitability of the method. Verification is the process to determine that a model correctly represents the developer's conceptual description. It is used to decide whether the model was "built" correctly. Validation is the process to determine that a model is a suitable representation of the real world and is capable of reproducing phenomena of interest. It is used to decide whether the right model was "built."

This project was driven by the following objectives:

- Conduct a V&V study of the selected fire models using a consistent methodology (ASTM E 1355, *Evaluating the Predictive Capability of Deterministic Fire Models* [4]) for NPP fire protection applications.

- Quantify predictive capabilities of fire models to the extent that can be supported by comparison with applicable and available fire experiment data.

This study evaluated the following five fire modeling tools:

(1) NRC's Fire Dynamics Tools (FDT^S) (documented in Volume 3 of this report)

(2) EPRI's Fire-Induced Vulnerability Evaluation, Revision 1 (FIVE-Rev1) (documented in Volume 4 of this report)

(3) National Institute of Standards and Technology's (NIST) Consolidated Model of Fire Growth and Smoke Transport (CFAST) (documented in Volume 5 of this report)

(4) Electricité de France's (EdF) MAGIC code (documented in Volume 6 of this report)

(5) NIST's Fire Dynamics Simulator (FDS) (documented in Volume 7 of this report).

1.2.2 Approach

This program follows the guidelines of ASTM E 1355, *Evaluating the Predictive Capability of Deterministic Fire Models* [4], which ASTM International distributes as a guide to evaluate fire models. That standard identifies four steps in the evaluation of predictive fire models:

(1) Define the model and scenarios or phenomena for which the evaluation is to be conducted.

(2) Assess the appropriateness of the theoretical basis and assumptions used in the model.

(3) Assess the mathematical and numerical robustness of the model.

(4) Validate the model by quantifying the uncertainty and relative difference[1] of the model results in predicting the course of events for specific NPP fire scenarios.

Traditionally, a V&V study reports on the comparison of model results with experimental data from a test series and, as such, the V&V of the fire model is for the specific tested fire scenarios. V&V studies for the selected fire models do already exist, but it is necessary to investigate the technical issues specific to the use of these models in NPP fire modeling applications.

In order to accomplish the ASTM E 1355 objectives, the following approach was developed and implemented in this study:

(1) **Define a list of typical NPP fire scenarios.**

(2) **Select test series from which experimental data will be used to perform the quantitative validation.**

(3) **Select and describe the fire models for which an evaluation can be conducted.**

(4) **Define fire modeling parameters.**

(5) **Conduct the quantitative validation study for each fire modeling tool.**

(6) **Report validation results.**

These steps are described in more detail in Chapter 2. This approach provides a roadmap to model users and developers for conducting a V&V based on ASTM E 1355.

The scope of this V&V study is limited to the capabilities of the selected fire models. As such, certain potential fire scenarios in NPP fire modeling applications do not fall within the capabilities of these fire models and, therefore, are not covered by this study. Examples of such fire scenarios include high-energy arcing faults and fire propagation between control panels [5, Section 7.2.2]. It is the user's responsibility to determine whether a model can be applied to each specific fire scenario.

1.3 Report Structure

This report is presented in seven volumes:

- Volume 1, "Main Report," provides general background information, programmatic and technical overviews, project results, insights, and conclusions. The description of the typical commercial NPP fire scenarios is contained in Section 2 of Volume 1.

---

1   See Section 2.5 for the specific definition of relative difference

- Volume 2 presents a summary description of the fire tests and estimates of experimental uncertainty used in this V&V study.

- Volumes 3 through 7 provide detailed discussions of the V&V of the FDT^s, FIVE-Rev1, CFAST, MAGIC, and FDS fire models. Each report follows the guidelines provided by ASTM E 1355 and contains the following chapters:

  - Chapter 1, Introduction

  - Chapter 2, Model Definition, briefly describes the fire model.

  - Chapter 3, Theoretical Bases for the Model, includes theoretical descriptions of the fire model. In addition, this chapter provides a literature review and discusses the capabilities, limitations, and range of applications of the model.

  - Chapter 4, Mathematical and Numerical Robustness, discusses the mathematical and numerical robustness of the fire model.

  - Chapter 5, Model Sensitivity, presents the results of sensitivity analyses conducted for the fire model. In general, the sensitivity analysis evaluates model variations from a base case scenario, as they are affected by changes in the input parameters.

  - Chapter 6, Model Validation, documents the methodology and results of the V&V study.

  - Chapter 7, References

  - Appendix A, Technical Details for Validation Study

  - Appendix B, Input Files

2
TECHNICAL APPROACH

ASTM E 1355 establishes a process for conducting a V&V study of a fire model. In general, the process can be summarized in the following tasks:

- **Model and scenario definition** documents the model and the scenarios or phenomena of interest for the V&V study.

- **Description of the theoretical basis for the model** documents a detailed technical description of the thermo-physical processes addressed by the fire model.

- **Mathematical and numerical robustness** documents an evaluation of the numerical implementation of the model.

- **Model sensitivity** documents a sensitivity analysis of the model.

- **Model evaluation** documents the results of the validation study.

There is, however, a technical challenge in implementing these tasks. Specifically, the universe of fire scenarios in commercial NPPs is large and diverse. Also, scenarios may have characteristics or attributes that either cannot be modeled using state-of-the-art computational fire models, and/or no experimental data is available to support a V&V study of that particular characteristic or attribute. Improvements in these two specific limitations — limited fire modeling capabilities and/or insufficient experimental data — are needed.

In order to address these challenges, and still perform a V&V study consistent with ASTM E 1355, the following approach has been selected:

(1) **Define a list of typical NPP fire scenarios.** This list of fire scenarios is intended to be a reflection of the wide range of fire scenarios found in NPPs (i.e., the scope of scenarios for which models would need validation). In the context of this V&V study, the list of scenarios attempts to capture all the potential fire scenarios and the resulting conditions that could be postulated in practical applications. However, some conditions in these scenarios cannot be predicted with available models or do not have any available experimental data to support a quantitative model evaluation. Scenarios are listed and described in Section 2.1.

(2) **Select test series from which experimental data will be used to perform the quantitative validation.** The selected test series reflects some of the characteristics of the fire scenarios included in the list described in item 1 above. The selected tests are listed in Table 2.2.

(3) **Select and describe the fire models for which an evaluation can be conducted.** Consistent with ASTM E 1355, the description of the selected fire models includes a review of the theoretical basis and fundamental assumptions, an assessment of the mathematical and numerical robustness, and a sensitivity analysis, as well as validation with experimental data. As suggested earlier, (1) not all the predictive capabilities of each model have been subjected to the V&V process, and (2) not all the fire-generated conditions in the library of fire

scenarios can be predicted with the capabilities of state-of-the-art models. Selected models are listed in Section 2.3.

(4) **Define fire modeling parameters**. It is necessary to identify fire-generated conditions for which a generic validation study can be conducted given the available models and experimental data. Based on the NPP fire scenarios, capabilities of the selected fire models, and the available experimental data, the project team identified 13 fire modeling parameters for which a quantitative validation study can be conducted. The fire modeling parameters are described in Section 2.4.

(5) **Conduct the quantitative validation study for each fire modeling tool**. The quantitative validation studies are conducted by comparing experimental data with fire modeling tool predictions.

(6) **Report validation results**. Results from the quantitative validation study are reported as relative differences for peak experimental measurements and model predictions, as well as graphical comparisons between experimental measurements and model predictions.

Figure 2-1 graphically represents this approach. The following sections describe the steps of this approach in greater detail. This approach is documented and implemented in the individual volumes using the data from Volume 2. It can be used as a roadmap to model users and developers for conducting a V&V for models other than those included in this study.

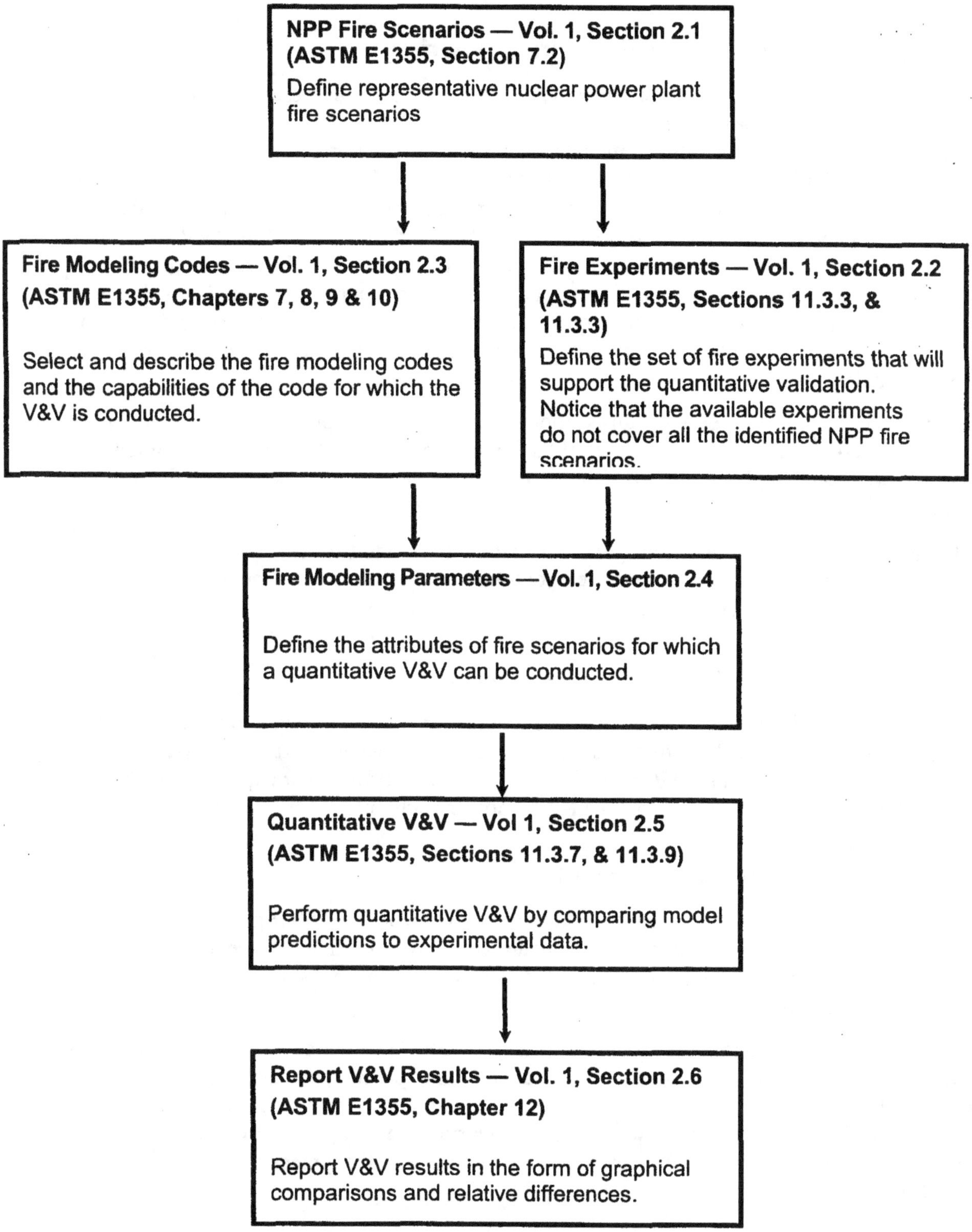

**NPP Fire Scenarios — Vol. 1, Section 2.1
(ASTM E1355, Section 7.2)**
Define representative nuclear power plant fire scenarios

**Fire Modeling Codes — Vol. 1, Section 2.3
(ASTM E1355, Chapters 7, 8, 9 & 10)**

Select and describe the fire modeling codes and the capabilities of the code for which the V&V is conducted.

**Fire Experiments — Vol. 1, Section 2.2
(ASTM E1355, Sections 11.3.3, &
11.3.3)**
Define the set of fire experiments that will support the quantitative validation. Notice that the available experiments do not cover all the identified NPP fire scenarios.

Fire Modeling Parameters — Vol. 1, Section 2.4

Define the attributes of fire scenarios for which a quantitative V&V can be conducted.

**Quantitative V&V — Vol 1, Section 2.5
(ASTM E1355, Sections 11.3.7, & 11.3.9)**

Perform quantitative V&V by comparing model predictions to experimental data.

**Report V&V Results — Vol. 1, Section 2.6
(ASTM E1355, Chapter 12)**

Report V&V results in the form of graphical comparisons and relative differences.

**Figure 2-1: Overview of the Approach for V&V Study of Selected Fire Models
for Nuclear Power Plant Applications**

2.1 Library of Nuclear Power Plant Fire Scenarios

To conduct the V&V study in accordance with ASTM E 1355, it is necessary to define the scenarios or phenomena of interest to evaluate each model. For the purpose of the V&V study, a fire scenario definition should include a complete description of the phenomena of interest in the evaluation to facilitate appropriate application of the model. As mentioned in the introduction to this chapter, this list of phenomena from the fire scenarios is intended to reflect the collection of phenomena from fire scenarios found in NPPs. In the context of this V&V study, the list of scenarios captures all the phenomena of interest that would be predicted by some fire models, but may or may not have experimental data to support a quantitative model evaluation. Consequently, this V&V study provides a quantitative evaluation for the phenomena defined by the scenarios to the extent allowed by the available experimental data and the capabilities of the selected models.

The list of fire scenarios presented in this section expands and/or modifies the list originally compiled and documented by EPRI as part of the development of its "Fire Modeling Guide for Nuclear Power Plant Applications" [5]. The basis for the selection of these fire scenarios is as follows:

- Review the range of possible configurations that contribute to fire scenarios in the U.S. commercial nuclear industry. The review focused on parameters considered important in the definition of fire scenarios.

- Identify potentially risk-significant fire scenarios through review of the Individual Plant Examination for External Events (IPEEE) submittals.

- Examine past industry experience with fire modeling in support of regulatory applications (other than IPEEE) to help define these fire scenarios. A questionnaire was prepared and distributed to all operating NPPs in the United States concerning their experience with fire modeling. Also, with support from the NRC, industry submittals were searched to identify the use of fire modeling.

Additional details are available in reference 5.

Further information on NPP fire scenarios is found in NUREG/CR-6850 [6]. This reference discusses risk methods that may be used to evaluate scenarios that can be outside the applicability of the fire modeling tools evaluated in this report. Such scenarios include high-energy arcing faults, main control board fires, and hydrogen fires.

The generic list of scenarios includes fires in the switchgear room (SWGR), cable spreading room (CSR), main control room (MCR), pump room, turbine building, multiple compartment (corridor) scenarios, multi-level building, containment (PWR), battery room, diesel generator room, computer room, and outdoors. The descriptions below of the fire scenario are examples of how and where a fire could start. The sources of fires described are representative of the typical configurations in most NPPs.

2.1.1 Switchgear Room

The SWGR is often an important area in a commercial NPP. A fire in a SWGR can have significant fire risk repercussions and, hence, the SWGR is one of the two plant locations that are most often identified as the top fire risk contributors in fire risk assessments performed under the IPEEE program[2]. The reason the SWGR can be essential to plant operation is that it typically contains equipment and circuits that provide the electrical power needed to operate and control the plant. This area also contains potential sources of high-energy arcing faults that may be located close to other safety-related equipment and/or circuits. Figure 2-2 graphically represents the SWGR fire room scenario. The most common fire source in this scenario may be an electrical cabinet. The size of the fire will depend on the type and amount of cables present in the cabinet and the ventilation conditions within the cabinet itself. Typically, these rooms have closed doors and are mechanically ventilated. Important targets may be cables in a cable tray located above the switchgear cabinet that are exposed to flame, plume or ceiling jet conditions or radiant heat flux from the fire. Targets may also be exposed to the hot gas layer thermal conditions.

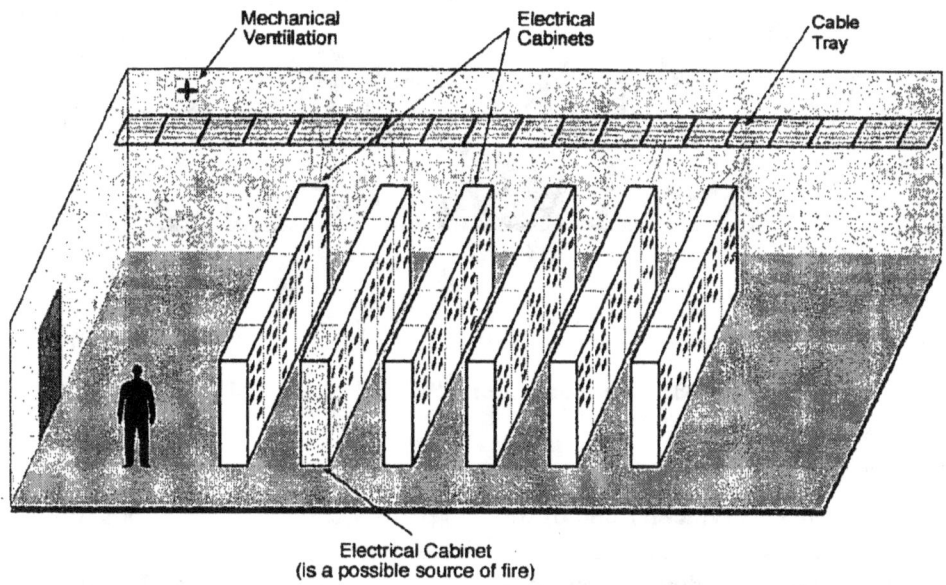

Figure 2-2: Switchgear Room Fire Scenario

2.1.2 Cable Spreading Room

The CSR is another critical location in a commercial NPP because it often contains redundant instrumentation and control circuits needed for plant operation. The CSR generally contains a high cable concentration (in cable trays and/or conduits), and fire propagation in open cable trays can be an important aspect of fire modeling. Some NPPs have areas called cable tunnels or cable lofts, which present similar challenges. These areas may also contain significant amounts of cables in trays or conduits and may contain redundant circuits. Figure 2-3 graphically represents the scenario. The source of a fire in this scenario may be transient combustibles, or electrical cable or cabinets. The ventilation conditions in the room may be mechanical ventilation, or possibly some level of natural ventilation via leakage around closed doors. Important targets

---

2   Individual NPP licensees conducted IPEEEs to assess the risks to the plant design from external events such as earthquakes, high winds, and internal fires.

may be cables in a cable tray located above the cabinet on fire that are exposed to plume or ceiling jets conditions, or an electrical cabinet or tray exposed to radiant heat flux from the fire. Targets may be also located in the hot gas layer and subjected to hot gas layer thermal conditions.

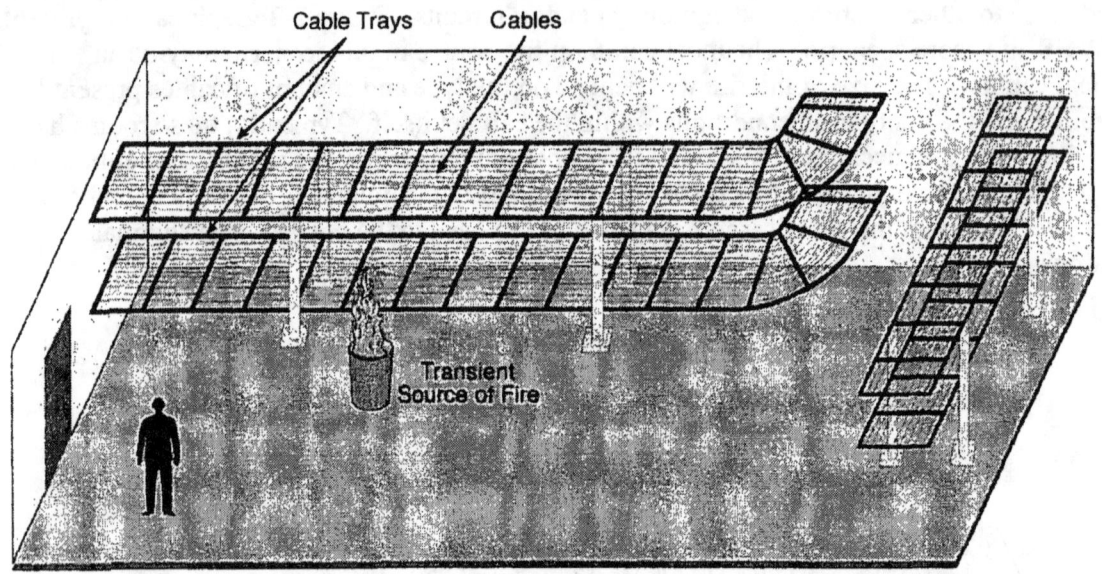

Figure 2-3: Cable Spreading Room Fire Scenario

2.1.3 Main Control Room

Like the SWGR, the MCR is typically one of the two plant locations that are most often identified as the top fire risk contributors in fire risk assessments performed under the IPEEE program. The MCR contains redundant instrumentation and control circuits that are critical to plant control and safe-shutdown. Analyses of fires in the MCR pose unique challenges, including timing of fire detection, smoke generation, migration, and habitability (including visibility and concentration of species); fire propagation within very large panels; and fire propagation between panels. It should also be noted that some NPPs have areas (i.e., a relay room, auxiliary equipment room, or remote shutdown panel) that are similar to MCRs, in that they contain redundant instrumentation and control circuits that are critical to plant control and safe-shutdown. However, such areas are not constantly manned like MCRs and may instead be equipped with automatic suppression systems. A fire in the MCR may lead to a situation where the reactor cannot be controlled due to damage to the instrumentation and control circuits there. Figure 2-4 graphically represents a MCR scenario. This scenario can apply to one or more unit NPP control rooms. The source of a fire in this scenario may be a control cabinet. The size of the fire will depend on the type and amount of cables within the cabinet, as well as cabinet ventilation and detection and suppression activities in the constantly-manned control room. The ventilation conditions in the room will be mechanical ventilation. Important targets are adjacent control cabinets exposed to radiant heat flux or flame impingement. Another important aspect of main control room fire scenarios is the habitability conditions in the room as the fire progresses. Habitability conditions refer to smoke concentration (which affects visibility, toxicity), heat flux from the hot gas layer, and room temperature. These conditions are important for determining when operators may need to leave the control room as a result of relatively high temperatures or low visibility.

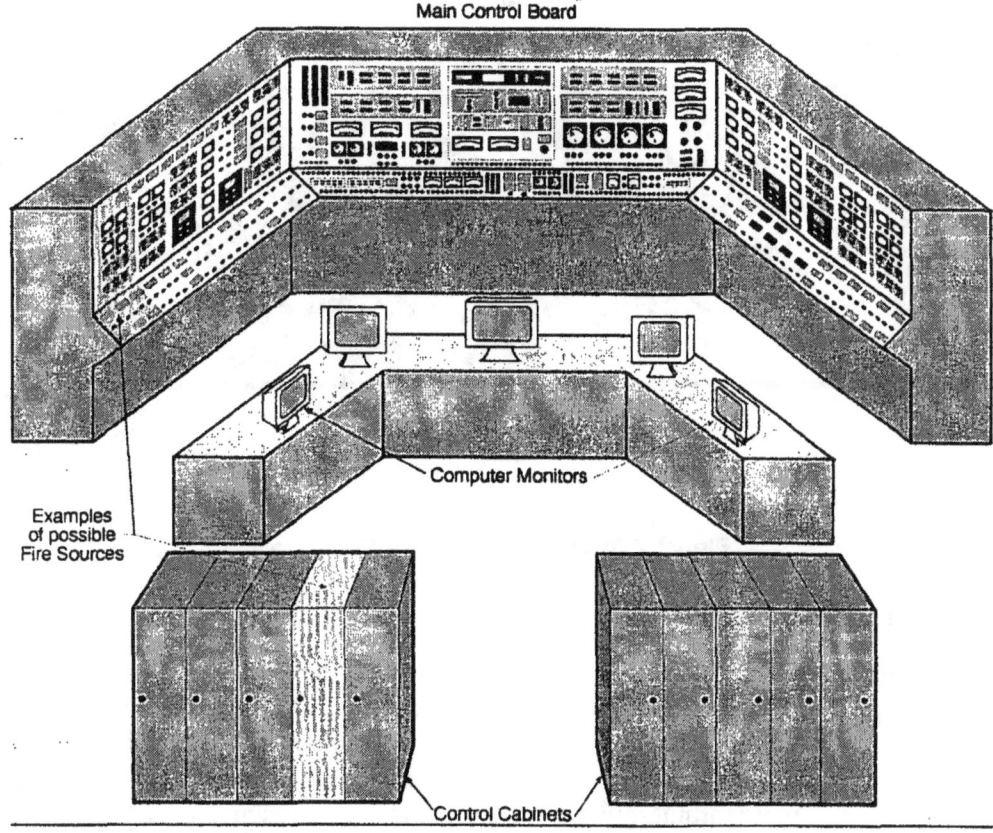

Figure 2-4: Main Control Room Fire Scenario

2.1.4 Pump Room

This location represents areas in a plant where a relatively large fire is possible in a small enclosure. However, not all pump rooms are small, since relatively large pumps can be found in large open areas such as turbine building elevations. Figure 2-5 graphically represents the scenario. The source of a fire in this scenario may be ignition of an oil pool spilled from a pump. The size of the fire will depend on the type and amount of oil spilled, as well as the area and depth of the pool itself. The growth of this fire typically will be fast, and depending on the size of the room, the fire could possibly generate flashover conditions that may challenge the walls and ceiling. The ventilation conditions in the room will be mechanical ventilation with leakage around closed doors. Targets of interest in these scenarios may be the walls and ceiling of the enclosure, which are fire barriers, as well as any other safety-related equipment and cables located in the room or area. These targets may be exposed to direct flame impingement or flame radiation or plume, ceiling jet, or hot gas layer conditions.

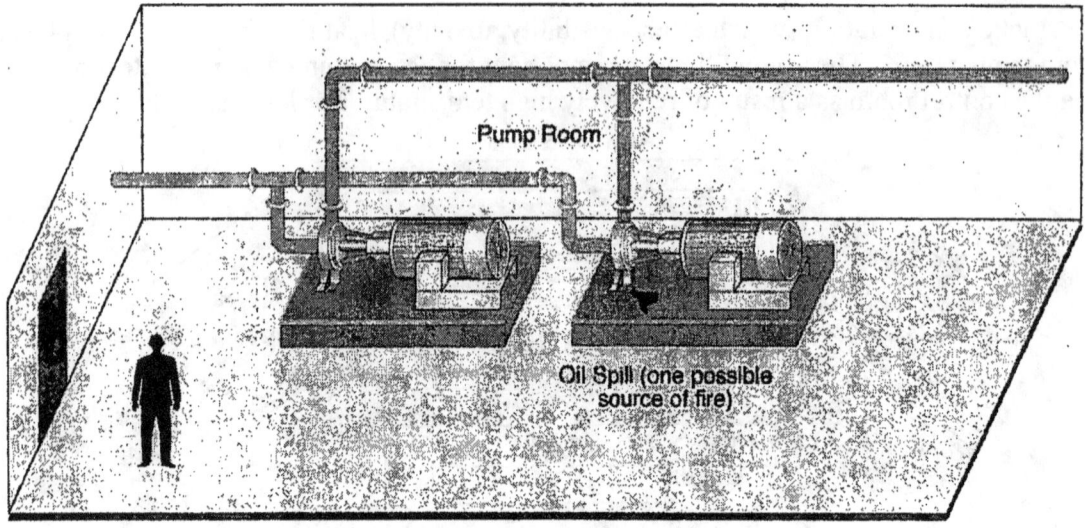

Figure 2-5: Pump Room Fire Scenario

2.1.5 Turbine Building

A turbine building is usually a multi-level enclosure[3], in which the top level is commonly referred to as the turbine operating deck. A fire scenario on the turbine deck was selected to examine large (e.g., turbine lube oil) or small (e.g., transient or panel) fires in large enclosures with high ceilings. A multi-level turbine building fire is described in Section 2.1.7. The scenario can apply to buildings with one or more turbines. Figure 2-6 graphically represents the scenario. The source of a fire in this scenario may be ignition of an oil pool spilled from one of the turbines. The size of the fire will depend on the type and amount of oil spilled, as well as the area and depth of the pool itself. The growth of this fire will be fast. Other sources of fire in the turbine building could be electrical fires (e.g. high energy arching faults), transformer or switchgear fires, and hydrogen fires. The ventilation conditions will be natural ventilation, with many open shafts (e.g. open equipment hatches), doors and windows. There may also be mechanical ventilation using roof-mounted exhaust fans and/or mechanical supply. Targets of interest in these scenarios may be structural steel members and fire barriers, as well as any other safety-related equipment and cables located in the area and exposed to the fire. Fire conditions affecting the targets may include direct flame impingement, fire plume conditions, or flame radiation.

---

3   Some NPPs (typically in warmer climates) do not have a turbine building, and the main turbine is open to the elements.

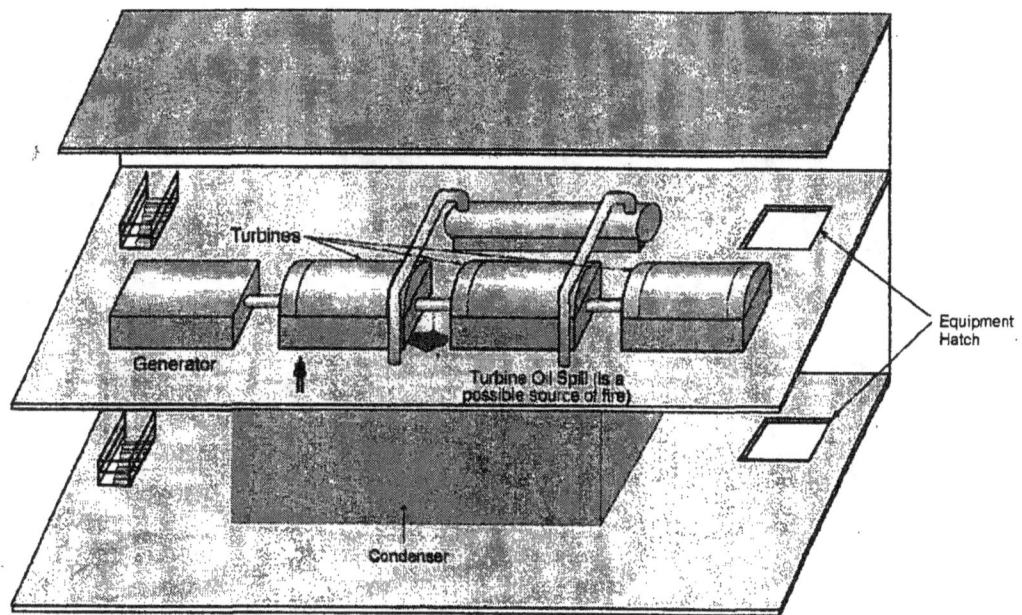

Figure 2-6: Turbine Building Fire Scenario

2.1.6 Multi-Compartment Corridor

Many commercial NPPs have enclosures with multiple compartments that open into a common space or corridor. The significance of these enclosures in terms of fire safety varies from plant-to-plant because they house various mechanical, electrical, waste treatment, or other equipment and/or circuits. Figure 2-7 graphically represents this scenario, which consists of a fire in one compartment affecting targets in an adjacent compartment. The multi-compartment corridor considered in this scenario consists of interconnected rooms and corridors in the same level. These geometries may have soffits between the connecting rooms. The source of a fire in this scenario may be ignition of an oil pool spilled from a pump in one of the adjacent rooms. The size of the fire will depend on the type and amount of oil spilled, as well as the area and depth of the pool itself. The growth of this fire typically will be fast and, depending on the size of the room, the fire could potentially generate flashover conditions in the room of origin, and fire effluent may spill out and effect targets in adjacent rooms. The ventilation conditions will be natural ventilation, with leakage paths between compartments around normally closed doors. There may also be mechanical ventilation using both injection and extraction systems. Targets of interest in these scenarios are often safety-related equipment and cables located in the corridor outside the room of fire origin, or an adjacent room. These targets will be subjected to smoke flows migrating out of the room of fire origin.

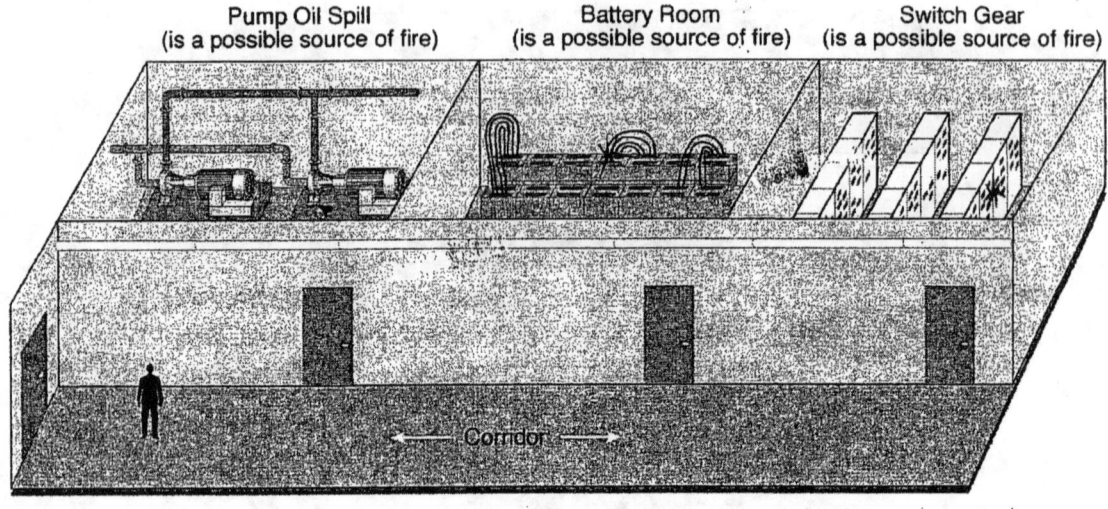

Figure 2-7: Multi-Compartment Corridor Fire Scenario

2.1.7 Multi-Level Building

A typical NPP has locations where multiple elevations in the same building are separated by partial floors/ceilings, open hatches, or staircases. Typical examples include turbine buildings, pressurized-water reactor (PWR) auxiliary buildings, and boiling-water reactor (BWR) buildings. The multi-level turbine building in this scenario is a three-level space that includes the turbine deck. (Section 2.1.5 describes a turbine deck fire scenario.) This scenario consists of an oil spill fire affecting targets located on a different level. Figure 2-8 graphically represents the scenario, which can apply to turbine buildings with one or more units. The source of a fire in this scenario may be ignition of an oil pool spilled from an oil tank located under one of the turbine generators. The size of the fire will depend on the type and amount of oil spilled, as well as the area and depth of the pool itself. The energy and smoke created will flow through mezzanine opening between levels. The growth of this fire will be fast. The ventilation conditions will be natural ventilation via openings on the upper level. There may also be mechanical ventilation using roof-mounted exhaust fans and/or mechanical supply. Targets of interest in these scenarios may be cables in cable trays located on the upper levels.

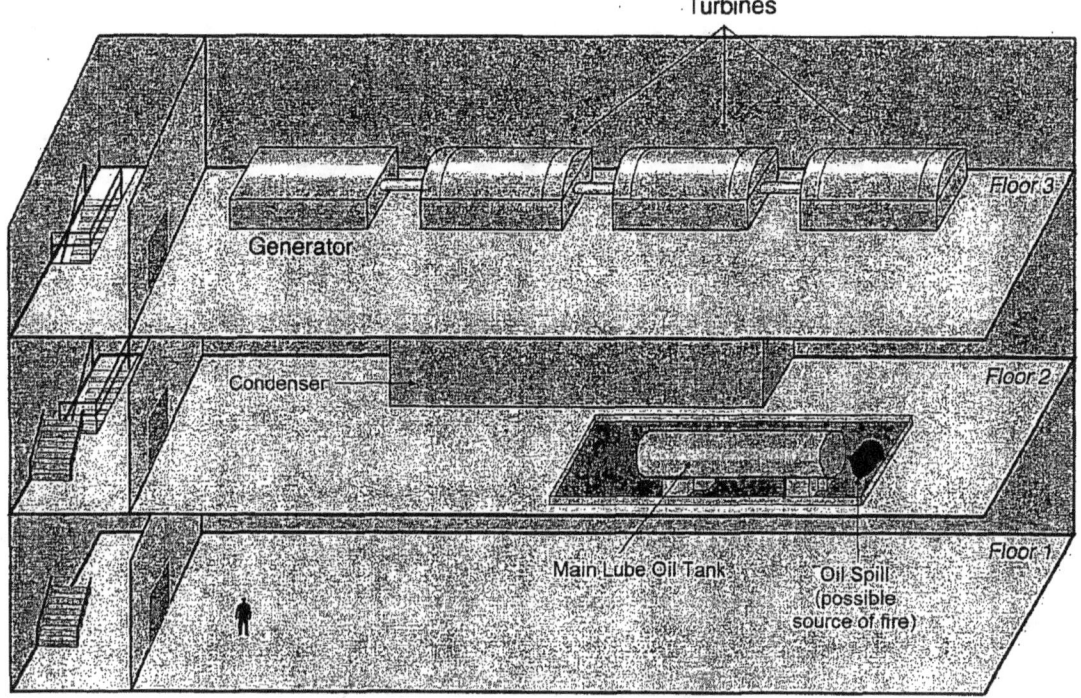

Figure 2-8: Multi-Level Building Fire Scenario

2.1.8 Containment Building (PWR)

The containment building in a PWR plant was selected because of its geometrical characteristics, which include cylindrical boundaries, a high domed ceiling, and a large volume. Figure 2-9 graphically represents the scenario, which involves an oil spill fire from the reactor coolant pump (RCP). The size of the fire will depend on the type and amount of oil spilled, as well as the area and depth of the pool itself. The containment in a PWR has internal air recirculation systems with cooling units. There is no fresh air added into the containment atmosphere during normal operation. The target of interest in this scenario is an elevated cable tray located outside the fire plume. These targets may be exposed to direct flame impingement or flame radiation or plume, ceiling jet or hot gas layer conditions.

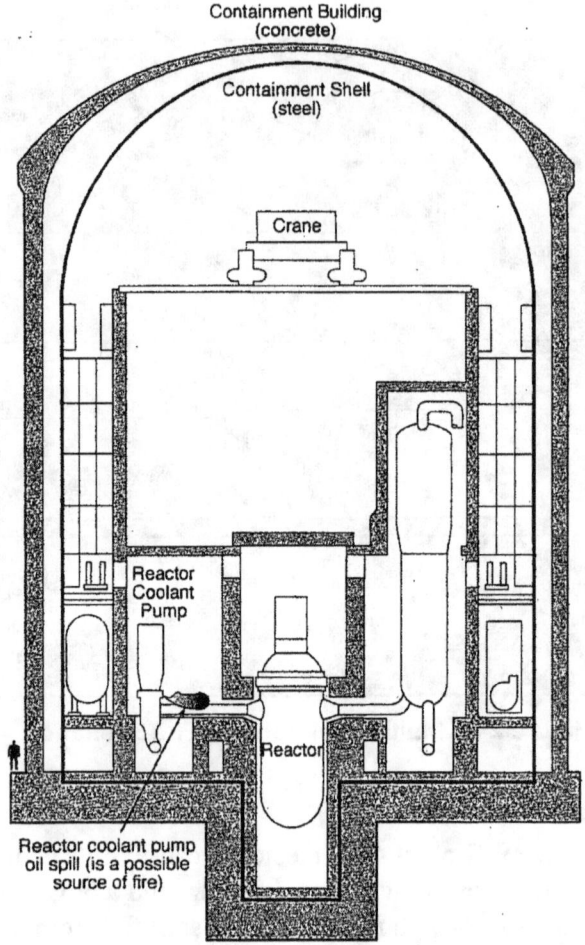

Figure 2-9: PWR Containment Building Scenario

2.1.9 Battery Room

Battery rooms are usually relatively small concrete rooms with two or more large banks of batteries. These rooms are kept closed and are typically free of transient combustibles and fixed ignition sources other than the batteries. EPRI's Fire Events Database suggests two types of scenarios: (1) explosion of the battery cells during the charging phase of the battery, and (2) fires in battery terminals as a result of defective or unsecured terminals. Rooms are usually mechanically ventilated. The targets of interest in this scenario are nearby cables and batteries. These targets may be exposed to direct flame impingement or flame radiation or plume, ceiling jet or hot gas layer conditions.

2.1.10 Diesel Generator Room

Diesel generator rooms house the standby diesel generator (SBDG) and associated electrical cabinets. This scenario consists of a fuel oil fire near the diesel generator. The size of the fire will depend on the type and amount of fuel oil spilled, as well as the area and depth of the pool itself. The growth of this fire will be fast. The ventilation conditions in the room will be mechanical ventilation with leakage around closed doors. Targets of interest in these scenarios may be cables located in the room exposed to HGL temperatures. These targets may be exposed to direct flame impingement or flame radiation or plume, ceiling jet, or hot gas layer conditions.

2.1.11 Computer or Relay Room

Computer rooms are typically located in close proximity to the main control rooms in NPPs. In addition to computers and other office equipment, some computer rooms may house control cabinets or banks of relay panels. The ignition source for this scenario is a transient combustible fire, namely a computer workstation. The size of the fire will depend on the amount and type of materials involved. The ventilation conditions will be mechanical ventilation. The targets of interest may be control cabinets or banks of relay panels or cables above the fire. These targets may be exposed to direct flame impingement or flame radiation or plume, ceiling jet, or hot gas layer conditions.

2.1.12 Outdoors

Outdoor fire scenarios can involve large oil-filled transformers or hydrogen tanks and can affect or propagate to nearby equipment. Nearby equipment can include other transformers, other electrical equipment, turbine building walls, etc. Considering that fires will be outdoors, fire conditions that could affect targets include flame radiation and exposure to fire plumes. In the case of transformers, fires can be attributable to oil leaks or spills or electrical faults. Consequences will depend on the type of fire analyzed (i.e., a "regular" fire vs. an explosion).

2.2 Fire Models

There are numerous fire models that have been developed and maintained by various organizations to predict fire-generated conditions. This study selects the following five of these fire models, which represent a wide range of capabilities and mathematical and computational sophistication:

- Two libraries of engineering calculations: FDT^s and FIVE-Rev1

- Two two-zone models: CFAST and MAGIC

- One field model: FDS

These particular models were chosen based on the fact that most of them have been used to calculate fire conditions in NPP fire protection applications, or were developed by stakeholders within the nuclear industry for NPP fire protection applications. FDS was chosen to represent the most complex types of models available for fire protection applications.

2.2.1 Libraries of Engineering Calculations: FDT^s and FIVE-Rev1

FDT^s is a library of engineering calculations (also referred to as hand calculations) in the form of Microsoft® Excel® spreadsheets. For the most part, the models in the FDT^s library are closed-form algebraic expressions programmed in spreadsheets to provide a user-friendly interface that reduces input and computational errors. Technical details concerning the engineering calculations and use of the spreadsheets are available in NUREG-1805 [7].

FIVE-Rev1 is another library of engineering calculations in the form of Microsoft® Excel® spreadsheets. Specifically, FIVE-Rev1 library consists of functions programmed in Visual Basic for Applications, which is the programming language within Excel®. Having the models programmed as Microsoft® Excel® functions allows the use of Excel® spreadsheets as the interface. Technical details concerning the engineering calculations and their use in the Excel® environment are available in EPRI TR-1002981 [5].

The FDTˢ and FIVE-Rev1 libraries include different models, and some of those models were not evaluated for this V&V study, because of applicability and a lack of experimental data. Table 2-1 shows the equations that were used from each library to evaluate fire scenario attributes. The fire scenario attributes that were selected for this study (see Section 2.4 for the description of these attributes) are those that are used in NPP fire modeling applications. For the most part, these selected models are included in both libraries.

Table 2-1: FDTˢ and FIVE-Rev1 Models for the Fire Scenario Attributes Selected for this V&V Study

Attribute (See Section 2.4)	FDTˢ	FIVE-Rev1
1. Hot gas layer temperature	MQH, FPA, Beyler, Beyler & Deal	MQH, FPA
2. Hot gas layer height	Yamana & Tanaka	No model
3. Ceiling jet temperature	No model	Alpert's ceiling jet temperature correlation
4. Plume temperature	Heskestad's plume temperature correlation	Heskestad's plume temperature correlation
5. Flame height	Heskestad's flame height correlation	Heskestad's flame height correlation
6. Radiated heat flux to targets	Point source flame radiation model, Solid flame model	Point source flame radiation model

2.2.2 Two-Zone Fire Models: CFAST and MAGIC

Fire modeling programs that were developed under the assumption that a fire will generate two distinct zones with uniform thermal properties are referred to as two-zone models. This V&V study evaluated two two-zone models, CFAST and MAGIC.

CFAST [8] is a two-zone fire model that predicts the environment that arises within compartments as a result of a fire prescribed by the user. CFAST was developed and is maintained primarily by the Fire Research Division of the National Institute of Standards and Technology (NIST). In terms of modeling capabilities, CFAST provides the average temperatures of the upper and lower gas layers within each compartment; flame height; ceiling, wall, and floor temperatures within each compartment; flow through vents and openings; visible smoke and gas species concentrations within each layer; target temperatures; heat transfer to targets; sprinkler activation time; and the impact of sprinklers on the fire heat release rate (HRR).

MAGIC [9] is a two-zone fire model developed and maintained by Electricité de France (EdF). It is available through EPRI to its members. In terms of modeling capabilities, MAGIC predicts (1) environmental conditions in the room (such as hot gas layer temperature, and oxygen/smoke concentrations), (2) heat transfer-related outputs to walls and targets (such as incident convective, radiated, and total heat fluxes), (3) fire intensity and flame height, and (4) flow velocities through vents and openings.

2.2.3 Field Fire Model: FDS

FDS [10] is a computational fluid dynamics (CFD) model of fire-driven fluid flow. The model numerically solves a form of the Navier-Stokes equations appropriate for low-speed, thermally driven flow, with an emphasis on smoke and heat transport from fires. The partial derivatives of the equations for conservation of mass, momentum, and energy are approximated as finite differences, and the solution is updated in time on a three-dimensional, rectilinear grid. Thermal radiation is computed using a finite volume technique on the same grid as the flow solver. Lagrangian particles are used to simulate smoke movement and sprinkler discharge.

FDS computes the temperature, density, pressure, velocity, and chemical composition within each numerical grid cell at each discrete time step. There are typically hundreds of thousands to several million-grid cells, and thousands to hundreds of thousands of time steps. In addition, FDS computes the temperature, heat flux, mass loss rate, and various other quantities at solid surfaces.

2.3 Experimental Data

This section provides a general overview of the test series and experiments selected for this study. Volume 2 augments this overview by providing detailed descriptions of these experiments. Some test series included many experiments, from which only a few were chosen for this V&V study. One overriding reason for this is that the sheer amount of data that is generated and must be processed can be overwhelming, so limiting the number of experiments to consider was necessary. The experiments within the test series that were chosen are representative of the overall series of tests, as well as representative of the fire scenarios in NPPs listed above. Volume 2, Section 1.1, has a more complete explanation for the selection of the experiments.

2.3.1 Factory Mutual & Sandia National Laboratories (FM/SNL) Test Series

A series of fire tests was conducted at Sandia National Laboratories (SNL) under the sponsorship of the NRC in the mid-1980s. Specifically, tests were conducted using simple gas-fired burner, heptane pool, methanol pool, and solid polymethyl-methacrylate (PMMA) fires. Four of these tests were conducted with a full-scale control room mockup in place. Parameters varied during testing were fire intensity, enclosure ventilation rate, and fire location. The primary purpose at the time of these tests was to provide data for use in validating computer fire environment simulation models that would subsequently be used in analyzing NPP enclosure fire scenarios, specifically MCR scenarios.

These tests were conducted in an enclosure measuring 18.3 m x 12.2 m x 6.1 m (60 ft x 40 ft x 20 ft), which was constructed at the Factory Mutual Research Corporation fire test facility in Rhode Island. All of the tests utilized forced ventilation conditions. The ventilation system was designed to simulate typical NPP installation practices and ventilation rates.

NUREG/CR-4681 [11] provides a detailed description of the FM/SNL test series, including the types and location of measurement devices as well as some results. Additional results are reported in NUREG/CR-5384 [12].

This study used data from only the three tests with data reported in NUREG/CR-4681 (namely FM/SNL Tests 4, 5, and 21). For all three tests, the fire source was a propylene gas-fired burner with a diameter of approximately 0.9 m (2.95 ft), with its rim located approximately 0.1 m (0.33 ft) above the floor. For FM/SNL Tests 4 and 5, the burner was centered along the longitudinal axis centerline, 6.1 m (20 ft) laterally from the nearest wall. For FM/SNL Test 21, the burner was placed within simulated benchboard electrical cabinets.

2.3.2 *The National Bureau of Standards (NBS) Test Series*

A total of 45 tests representing 9 different sets of experiments, with multiple replicates of each set, were conducted in a three-room suite at NBS that is described in detail in NBSIR 88-3752 [13]. These tests were conducted in the mid-1980s as well. NBS is now known as NIST. The suite consisted of two relatively small rooms, designated here as Rooms 1 and 3, which were connected via doorways and short connecting passageways to a relatively long corridor, designated as Room 2. Rooms 1 and 3 opened only onto the corridor (Room 2) via doorways; they did not open to the external environment other than through normal construction leakage paths. The corridor had a doorway to the external environment, as well as doorways to Rooms 1 and 3. The fire source, a gas-fired burner, was located against the rear wall of Room 1. The following parameters were varied in the 9 different sets of experiments:

- fire size, including nominal 100, 300, and 500 kW fires

- door positions, including open and closed doors between the corridor and Room 3, as well as between the corridor and the external environment

In this study, the tests designated as Sets 1, 2, and 4 in NBSIR 88-3752 are used for comparison. All three of these sets had a fire source intensity of 100 kW, but the sets differed based on door position. Specifically, the door between Rooms 1 and 2 was open for all three sets. However, for Set 1, the door between Room 2 and the external environment was open (providing a source of fresh air to the suite), while the door between Rooms 2 and 3 was closed (effectively isolating Room 3 from this test). By contrast, for Set 2, the door between Room 2 and the external environment was closed, as was the door between Rooms 2 and 3 (again isolating Room 3). For Set 4, the door between Room 2 and the external environment was open (again providing a source of fresh air to the suite), as was the door between Rooms 2 and 3.

Experimental data used for these comparisons was obtained in electronic format from NIST. These data were converted to spreadsheet format for tests designated as MV100A through MV100AB. Average values for the nine data sets were also converted to spreadsheet format, but were not used for these comparisons. Rather, an exemplar test was selected from each data set for comparison purposes. Specifically, Test MV100A was used for Set 1, Test MV100O was used for Set 2, and Test MV100Z was used for Set 4. The selected data are also available in EPRI's Fire Modeling Code Comparison TR-108875 [14].

2.3.3 The International Collaborative Fire Model Project (ICFMP) Benchmark Exercise Test Series

To date, four full-scale fire test series have been completed as part of the ICFMP. The ICFMP is a separate, but related, project designed to conduct validation studies of fire models from around the world. ICFMP participants conduct a series of experiments and provide the data to other participants in order to compare fire model outputs to experimental data. These test series are referred to as benchmark exercises (BEs). This V&V study includes experimental data from BEs #2, 3, 4, and 5. A brief description of each follows:

- BE #2 [15]: These tests were conducted in the late-1990s. The ICFMP objective of BE #2 was to examine scenarios that are more challenging for zone models. In particular, these scenarios included fires in larger room volumes that are representative of turbine halls in NPPs. The tests were conducted inside the VTT Fire Test Hall, which has dimensions of 19 m high x 27 m long x 14 m wide (62.3 ft x 88.6 ft x 45.9 ft). Each case involved a single heptane pool fire, ranging from 2 MW to 4 MW.

- BE #3 [16]: This ICFMP exercise comprised a series of 15 large-scale fire tests, sponsored in part by the NRC, that were performed at NIST between June 5 and 20, 2003. These tests consisted of 350 kW, 1.0 MW, and 2 MW fires in a marinite room with dimensions of 21.7 m x 7.15 m x 3.7 m (71.2 ft x 23.5 ft x 12.1 ft). The room had one door with dimensions of 2 m x 2 m (6.6 ft x 6.6 ft), and a mechanical air injection and extraction system. Ventilation conditions and fire size were varied among the 15 tests. The numerous experimental measurements included temperatures in gas layers and surfaces, heat fluxes, and gas velocities, among others.

- BE #4 [17]: This test series was conducted at the Institut für Baustoffe, Massivbau Und Brandschutz (iBMB), in Germany in 2003 and 2004. Each of these tests simulated a relatively large fire in a relatively small concrete room. Only one test from this series was selected for this study.

- BE #5 [18]: This exercise, which was conducted at the iBMB in Germany in 2003 and 2004, consisted of four large-scale tests inside the same concrete enclosure as BE #4 with realistically routed cable trays. Only one test was selected for this study.

2.4 Selection of Fire Modeling Parameters

A complete V&V of a particular model for a given NPP fire scenario may not be possible because (1) the availability of experimental data is limited, and/or (2) a selected tool has limited modeling capabilities. Therefore, this implementation of ASTM E 1355 is intended to benefit as much as possible from the available data to establish the modeling capabilities and limitations of the selected fire models in typical NPP fire scenarios.

An important consideration in evaluating the capabilities of fire models in NPPs is the range of fire scenarios for which the models may be used. From a fire modeling perspective, most NPPs have similar configurations and fire hazards. That is, fire scenarios are characterized by similar attributes. Consequently, the V&V project team developed the following list of typical NPP fire modeling parameters, which provided the basic framework for conducting the V&V study and classifying the quantitative results:

(1) **Hot gas layer temperature**: The hot gas layer temperature is particularly important in NPP fire scenarios because it can provide an indication of target damage away from the ignition source. Models predict the increase in environmental temperature attributable to the energy released by a fire in a volume. However, different models define this volume in different ways. In the FDT^s and FIVE-Rev1 models available for predicting hot gas layer temperature, the assumed volume is the volume of the upper layer of the room and the output is a uniform upper layer temperature. In the CFAST and MAGIC two-zone models, the room is divided into upper and lower control volumes. Thus, the hot gas layer temperature output from these two-zone models is a uniform temperature in the upper control volume (which is referred to as the hot gas layer because it accumulates hot gases that are transported to the upper part of the room by the fire plume). Finally, in the FDS field model, the room is divided into numerous control volumes. Thus, FDS can provide outputs for the average temperature of the control volumes in the upper layer of the computational domain, as determined by a reduction of temperature profile data.

(2) **Hot gas layer height**: The height of the hot gas layer is also important in NPP fire scenarios because it indicates whether a given target is immersed in and affected by hot gas layer temperatures. The concept of hot gas layer height is most relevant in two-zone models (CFAST and MAGIC), in which this attribute defines the interface between the upper and lower control volumes. The FDS field model also provides the hot gas layer height output, which is calculated from the temperature profile within the height of the room. In addition, the FDT^s library includes one model that predicts hot gas layer height before the layer reaches a vent, assuming steady-state fire conditions in the room. However, because of the assumptions of this model and the algorithm used to process the experimental data, it does not apply to or could not be compared to the tests series included in this study. The FIVE-Rev1 library does not have a model for predicting hot gas layer height in the scenarios evaluated in this V&V study.

(3) **Ceiling jet temperature**: The ceiling jet is the shallow layer of hot gases that spreads radially below the ceiling as the fire plume flow impinges on it. This layer of hot gases has a distinct temperature that is higher than the temperature associated with the hot gas layer. This attribute is important in NPP fire scenarios that subject targets to unobstructed ceiling jet gases. The FIVE-Rev1 and MAGIC models calculate ceiling jet temperature using a semi-empirical correlation, and the ceiling jet temperature can be obtained from the FDS model

by inspecting the temperature profile in the pre-defined grid. The FDT^s library does not include a model for calculating ceiling jet temperature, and the CFAST model does not provide ceiling jet temperature as a direct output.

(4) **Plume temperature**: The fire plume is the buoyant flow rising above the ignition source, which carries the hot gases that ultimately accumulate in the upper part of a room to form the hot gas layer. The plume is characterized by a distinct temperature profile, which is expected to be higher than the ceiling jet and hot gas layer. This attribute is particularly important in NPP fires because of the numerous postulated scenarios that involve targets directly above a potential fire source. Models in the FDT^s and FIVE-Rev1 libraries predict fire plume temperatures using closed-form semi-empirical correlations, and the MAGIC model predicts plume temperatures in a similar manner. The plume temperature can also be obtained from the FDS model by inspecting the temperature profile in the pre-defined grid. The CFAST model does not provide plume temperatures as an output.

(5) **Flame height**: The height of the flame is important in those NPP fire scenarios where targets are located close to the ignition source. Some of these scenarios subject the target to flame temperatures because the distance between the target and the ignition source is less than the predicted flame height. A typical example would be cable trays above an electrical cabinet. Models in the FDT^s and FIVE-Rev1 libraries predict flame height using a close form semi-empirical correlation. MAGIC and CFAST models also predict flame height in a similar manner. The FDS combustion model has the capability to calculate flame height.

(6) **Radiated heat flux to targets**: Radiation is an important mode of heat transfer in fire events. The modeling tools within the scope of this study address fire-induced thermal radiation (or radiated heat flux) with various levels of sophistication, from simply estimating flame radiation, to calculating radiation from different surfaces and gas layers in the computational domain. The FDT^s and FIVE-Rev1 libraries include models for calculating flame radiation at a specified distance from the flames. By contrast, CFAST, MAGIC, and FDS have sophisticated heat transfer models that account for radiation exchanges between room surfaces and the upper and lower gas layers. Therefore, the incident thermal radiation to which a given target is exposed is a result of the heat balance at the surface of the target (which includes all of the exchanges), as well as the thermal radiation received from the flames.

(7) **Total heat flux to targets**: In contrast to thermal radiation (or radiated heat flux), the total heat flux a target is subjected includes convective heat transfer. Convective heat transfer is a significant contributor to target heat-up in scenarios that involve targets in the hot gas layer, ceiling jet, or fire plume. The CFAST and MAGIC two-zone models and the FDS field model account for convection, although CFAST (in particular) does not model target heating in the plume and ceiling jet sub-layers. The heat transfer models in the FDT^s and FIVE-Rev1 libraries do not account for convective heat transfer.

(8) **Total heat flux to walls**: This attribute was included as a separate attribute in this V&V study in order to evaluate model capabilities to determine the incident heat flux to walls, floors, and ceilings, which includes the contributions of radiation and convection. Because the heat conducted through the walls, floors, and ceilings does not contribute to room heat-up, it can be an important factor in the heat balance in control volume(s) in contact with the

surfaces. Of the models within the scope of this study, only the CFAST and MAGIC two-zone models and the FDS field model calculate total heat flux to walls, floors, and ceilings.

(9) **Wall temperature**: This attribute was included as a separate attribute in this V&V study to evaluate model capabilities to determine the temperature of walls, floors, and ceilings. Of the models within the scope of this study, only the CFAST and MAGIC two-zone models and the FDS field model provide the temperatures of these surfaces as outputs, since such outputs are part of the calculations required to determine the heat losses through boundaries.

(10) **Target temperature**: The calculation of target temperature is perhaps the most common objective of fire modeling analyses. The calculation of target temperature involves an analysis of localized heat transfer at the surface of the target after determining the fire-induced conditions in the room. The CFAST and MAGIC two-zone models and the FDS field model calculate the surface temperature of the target as a function of time, and consider the heat conducted into the target material. By contrast, the available model in the FIVE-Rev1 library assumes a constant incident heat flux and a semi-infinite solid. The FDT^s library does not include a model for estimating target temperature.

(11) **Smoke concentration**: The smoke concentration can be an important attribute in NPP fire scenarios that involve rooms where operators may need to perform actions during a fire. This attribute specifically refers to soot concentration, which affects how far a person can see through the smoke (visibility). The CFAST and MAGIC two-zone models and the FDS field model calculate smoke concentration as a function of time. These models determine smoke concentration as the fire plume carries combustion products into the hot gas layer. The FDT^s and FIVE-Rev1 libraries do not contain direct outputs of smoke concentration.

(12) **Oxygen concentration**: Oxygen concentration is an important attribute potentially influencing the outcome of fires in NPPs because of the compartmentalized nature of NPPs. Oxygen concentration has a direct influence on the burning behavior of a fire, especially if the concentration is relatively low. The CFAST and MAGIC two-zone models calculate the oxygen concentration in the upper and lower layers, and the FDS model calculates the oxygen concentration in each control volume defined in the computational domain. The FDT^s and FIVE-Rev1 libraries do not include models for calculating oxygen concentration.

(13) **Room pressure**: Room pressure is a rarely used attribute in NPP fire modeling. It may be important when it contributes to smoke migration to adjacent compartments. CFAST, MAGIC and FDS calculate room pressure as they solve energy and mass balance equations in the control volume. FDT^s library has a model for a sealed compartment that is not validated in this study. FIVE-Rev1 library does not have correlation to calculate room pressure.

Table 2-2: Fire Modeling Attributes as Outputs

Fire Modeling Attributes	Fire Models				
	FDT^s	FIVE	CFAST	MAGIC	FDS
Hot Gas Layer Temperature	Yes	Yes	Yes	Yes	Yes
Hot Gas Layer Height	Yes[1]	No	Yes	Yes	Yes
Ceiling Jet Temperature	No	Yes	Yes	Yes	Yes
Plume Temperature	Yes	Yes	No	Yes	Yes
Flame Height	Yes	Yes	Yes	Yes	Yes
Radiated Heat Flux to Targets	Yes	Yes	Yes	Yes	Yes
Total Heat Flux to Targets	No	No	Yes	Yes	Yes
Total Heat Flux to Walls	No	No	Yes	Yes	Yes
Wall Temperature	No	No	Yes	Yes	Yes
Target Temperature	No	No	Yes	Yes	Yes
Smoke Concentration	No	No	Yes	Yes	Yes
Oxygen Concentration	No	No	Yes	Yes	Yes
Room Pressure	No	No	Yes	Yes	Yes

[1] This output was not evaluated because it was not applicable for experiments used in this study.

2.5 Quantitative Validation

In keeping with the guidance in the ASTM E 1355 and the objectives of this study, the following approach was used for quantification of the results of the validation of the selected fire models.

The numerical comparison between an experimental observation and a corresponding model prediction is referred to as "relative difference" throughout this report. Relative differences have been calculated for each of the attributes listed in Section 2.4 using point estimate peak values from fire experiments and model predictions. The following equation, which is described in reference 4, has been selected for relative difference calculations:

$$\varepsilon = \frac{\Delta M - \Delta E}{\Delta E} = \frac{(M_p - M_o) - (E_p - E_o)}{(E_p - E_o)}$$

where ΔM is the difference between the peak value (M_p) of the model prediction and the ambient value (M_o), and ΔE is the difference between the experimental observation (E_p) and the ambient value (E_o). In the context of this study, for the parameters Oxygen Concentration and HGL Height, the "peak" value is actually the minimum value.

Table 2-3 summarizes the fire experiments and instruments used for quantitative validation of the different fire modeling parameters. The limited amount of data in the table is a reflection of the experimental data sets used in this study and, in general, the availability of reliable data in the literature. Data for some of the parameters (i.e., smoke, compartment pressure, radiant and total heat flux, etc.) was not collected in all of the test series. Because the data used to evaluate each parameter is different, the generality of each of the results is correspondingly limited. If more data becomes available in the future, it should be used to update the results in this report series.

The graphical comparisons of measured and predicted fire-generated condition profiles and calculated relative differences for each of the five fire models are detailed in Appendix A to Volumes 2 through 6. This information is the basis for the conclusions summarized in this volume.

Table 2-3: Summary of the Fire Tests Used for Validation against Typical NPP Fire Scenario Attributes

Fire Modeling Parameters	Selected Test Series/Experiments/Sensors					
	ICFMP BE #3 — Tests 1-5, 7-10, 13-18	ICFMP BE #5 — Test 4	FM/SNL — Tests 4, 5, & 21	ICFMP BE #4 — Test 1	ICFMP BE #2 — Part I, Cases 1, 2, 3	NIST Multi-Room Tests — 100A, 100O, 100Z
1. HGL temperature	Vertical thermocouple arrays	Vertical thermocouple arrays	Vertical thermocouple arrays	Vertical thermocouple arrays	Vertical thermocouple arrays	Vertical thermocouple arrays
2. HGL height	Vertical thermocouple arrays	Vertical thermocouple arrays	Vertical thermocouple arrays	Vertical thermocouple arrays	Vertical thermocouple arrays	Vertical thermocouple arrays
3. Ceiling jet temperature	Thermocouple	NA	Thermocouple	NA	NA	NA
4. Plume temperature	NA	NA	Thermocouple	NA	Thermocouple	NA
5. Flame height	Pictures	No Data	No Data	No Data	Pictures	No Data
6. Radiant heat flux to target (cables)	Radiometers	No Data	No Data	No Data	No Data	No Data
7. Total heat flux to targets	Heat flux gauges	Heat flux gauges	No Data	Heat Flux Gauges	No Data	No Data
8. Total heat flux to walls	Heat flux gauges	No Data	No Data	No Data	No Data	No Data
9. Wall surface temperature	Thermocouples	Thermocouples	No Data	Thermocouples	No Data	No Data
10. Target (cable) surface temperature	Thermocouples	Thermocouples	No Data	Thermocouples	No Data	No Data

Fire Modeling Parameters	Selected Test Series/Experiments/Sensors						
	ICFMP BE #3	ICFMP BE #5	FM/SNL	ICFMP BE #4	ICFMP BE #2	NIST Multi-Room Tests
	Tests 1-5, 7-10, 13-18	Test 4	Tests 4,5, & 21	Test 1	Part I, Cases 1, 2, 3	100A, 100O, 100Z
11. Smoke concentration	Smoke obs./conc.	No Data	No Data	No Data	No Data	No Data
12. Oxygen concentration	Oxygen meter	Oxygen meter	No Data	Flawed Data	No Data	No Data
13. Room pressure	Pressure transducer	No Data	No Data	No Data	No Data	No Data

2.6 Reporting and Applicability of the Validation Results

Section 2.1 of this volume listed typical NPP fire scenarios. The validation results of this study are limited by the general characteristics of the fire experiments selected. Consequently, the validation results need to be identified as corresponding to specific NPP fire scenarios to determine its applicability.

In general, the use of the quantitative results of this validation in support of fire modeling requires the following two steps:

1. *Applicability of V&V Results*: First, the user needs to assess the applicability of the validation results for the scenario under consideration.

2. *Characterization of fire model predictions based on the V&V results*: Once the user determines the applicability of the validation, the user must determine the level of confidence in the model prediction based on the quantitative results of this validation.

2.6.1 Applicability of the V&V Results

The following is one method that may be used to determine applicability of these validation results to other specific NPP fire scenarios. The description of this method is reported here to demonstrate the rigor users should use in determining applicability of these validation results. Other methods may be appropriate.

The applicability of the validation results can be determined using normalized parameters traditionally used in fire modeling applications. Normalized parameters allow users to compare results from scenarios of different scales by normalizing physical characteristics of the scenario.

Table 2-4 lists selected normalized parameters that may be used to compare encountered scenarios with the experiments used in this validation study. It is intended to provide guidance on which groups to consider when evaluating a certain attribute based on the validation results. Table 2-5 lists the ranges of values for different physical characteristics and normalized parameters based on the experiments considered in this validation study.

The user could calculate the normalized parameters that are relevant to the fire scenario being evaluated. If the parameters fall within the ranges evaluated in this validation, then the results of this study offer appropriate validation for the scenario. If they fall outside the range, then a validation determination cannot be made based on the results in this study. For any given fire scenario, more than one normalized parameter may be necessary for determining applicability of the validation results.

Table 2-4: List of Selected Normalized Parameters for Application of the Validation Results to NPP Fire Scenarios

	Normalized Parameter	General Guidance[1]
Heat release rate & flame height [19]	$Q_d^* = \dfrac{\dot{Q}}{\rho_\infty c_p T_\infty \sqrt{gD} D^2}$,	This parameter may be used for determining if the heat release rate and flame height in the analyzed scenario is within the range of heat release rates and flame heights within the scope of this study.
Room/Target height [20]	$\dfrac{H}{D^*}$, where $D^* = \left(\dfrac{\dot{Q}}{\rho_\infty c_p T_\infty \sqrt{g}}\right)^{2/5}$	This parameter may be used for scenarios involving a target immersed in the fire plume. This parameter suggests if the target elevation above the base of the fire is within the range included in this study.
Ceiling jet radial distance	$\dfrac{r_{cj}}{H}$	This parameter may be used in scenarios involving a target in the ceiling jet. This parameter suggests if the horizontal radial distance from the target to the centerline of the plume is within the range included in this study.
Natural Ventilation [19]	$\phi = \dfrac{\dot{m}_F/\dot{m}_{O_2}}{r} = \dfrac{\dot{Q}/\Delta H_c}{r} \Big/ 0.23 \times \dfrac{1}{2} A_o \sqrt{h_o}$	This parameter may be used for determining if the natural ventilation conditions in the room during the fire event is within the range of conditions included in this V&V report.
Mechanical Ventilation	$\phi = \dfrac{\dot{m}_F/\dot{m}_{O_2}}{r} = \dfrac{\dot{Q}/\Delta H_c}{r} \Big/ 0.23 \times \rho_{air} \dot{V}$	This parameter may be used for determining if the mechanical ventilation conditions in the room during the fire event is within the range of conditions included in this V&V report.
Room Size	$\dfrac{W}{H}$, or $\dfrac{L}{H}$, W and L are room width and length, respectively. H is room height.	This parameter may be used for scenarios in which the room size is an input to the fire model. These parameters suggest if the room size in the scenario under analysis is within the room sizes included in this V&V study.
Rad heat flux	$\left(\dfrac{D}{r}\right)^2$ where D is the diameter of the fire and r is the horizontal radial distance	This parameter may be used for scenarios involving radiated heat flux from flames. This parameter suggests if the horizontal radial distance from the fire to the target is within the range of distances included in this V&V study.

1. See detailed description of normalized parameters and examples for additional information

A detailed description of the normalized parameters follows.

Heat Release Rate (HRR or $\dot{Q}$): The most important parameter of any fire experiment is the HRR. In some cases, the fire model is used to predict the HRR. Here, however, the HRR is given, and the model is used to predict how the fire's energy is transported throughout the space of interest. A normalized quantity relating the HRR to the diameter of the fire, D, is commonly known as Q_d^*:

$$Q_d^* = \frac{\dot{Q}}{\rho_\infty c_p T_\infty \sqrt{gD}D^2}$$

Where: $\dot{Q}$ is the heat release rate (kW),

ρ_∞ is the ambient density (kg/m^3),

T_∞ is the ambient temperature (K),

c_p is the specific heat of ambient air(kJ/kg-K),

g is the acceleration of gravity (m/s^2), and

D is the diameter of the fire (m).

A large value of Q_d^* describes a fire whose energy output is relatively large compared to its physical diameter, like an oil well blowout fire. A low value describes a fire whose energy output is relatively small compared to its diameter, like a brush fire. Most conventional accidental fire scenarios have Q_d^* values on the order of 1. Its relevance to the current validation study is mainly in the assessment of flame height.

Example 1: An oil spill fire at floor level is postulated in a pump room. The scenario consists of 3.8 L (1 gal) of oil spilled in an area of 5.3 m^2 (57.0 ft^2). The target in this scenario is a cable crossing the room near the ceiling. The height of the pump room is 7 m (23.0 ft). Is the heat release rate calculated for this scenario within the ranges of heat release rates included in the V&V study? Assume that the heat of combustion for oil is 46,400 kJ/kg, and the mass loss rate is 0.036 kg/m^2-s (NUREG 1805, pp. 3–4).

The following steps are recommended to address this example:

1. Calculated the heat release rate of an 5.3 m^2 oil spill on fire:

   $\dot{Q} = m'' \Delta H_{c,eff} A_{dike} (1 - e^{-k\beta \cdot D}) = (0.036 kg/m^2 \cdot s)(46,400 KJ/kg)(5.3 m^2)(1 - e^{-(0.7*2.6)})$

   $\dot{Q} = 7416 kW$

2. Calculate Q_d^*

   $$Q_d^* = \frac{\dot{Q}}{\rho_\infty c_p T_\infty \sqrt{gD}D^2} = \frac{7416 kW}{(1.2 kg/m^3)(1 kJ/kg \cdot K)(293K)\sqrt{(9.81 m/s^2)(2.6m)}(2.6m)^2}$$

   $Q_d^* = 0.62$

3. Compare Q_d^* with the corresponding values listed in Table 2-5.

   The value of 0.62 is within the range of 0.4 – 2.4 listed in Table 2-5 for $\dot{Q}_d^*$.

Room/Target Height: The quantities Q_d^* and D^* relate the fire's HRR to its physical dimensions. The height of the compartment relative to D^* indicates the relative importance of the fire plume to the overall transport of the hot gases. Much of the mixing of fresh air and combustion products takes place within the plume, and this dilution of the smoke and the decrease in the gas temperature ultimately determines the hot gas layer temperature. Thus, the parameter H/D^* can be used to assess the importance of the plume relative to other features of the fire-driven flow, like the ceiling jet or doorway flow. In this normalized parameter, H is the target or ceiling elevation above the fire source and D* is defined as:

$$D^* = \left(\frac{\dot{Q}}{\rho_\infty c_p T_\infty \sqrt{g}} \right)^{2/5}$$

All variables are as defined above. The physical diameter of the fire is not always a well-defined property. A typical compartment fire may not have a well-defined diameter, whereas a circular pan filled with a burning liquid fuel has an obvious diameter. Fortunately, it is not the physical diameter of the fire that matters when assessing the "size" of the fire, but rather a characteristic diameter, D*.

In many instances, D^* is comparable to the physical diameter of the fire (in which case Q_d^* is on the order of 1). This group should be used when the fire scenario consists of a fire with a given diameter. A good example application is the evaluation of validation results for unconfined oil spill fire scenarios.

In summary, the H/D^* group is appropriate for evaluating applicability of validation results for scenarios involving targets inside the fire plume.

Example 2: Using the information from Example 1, and assuming the cable is inside the fire plume, is the target location in the fire plume within the range of locations considered in this V&V report?

The following steps are recommended to address this example:

1. Calculate D* using the heat release rate from Example 1.

$$D^* = \left(\frac{\dot{Q}}{\rho_\infty c_p T_\infty \sqrt{g}} \right)^{2/5} = \left(\frac{7416kW}{(1.2kg/m^3)(1kJ/kg \cdot K)(293K)\sqrt{9.81m/s^2}} \right)^{2/5} = 2.144m$$

2. Calculate H/D^*

$$\frac{H}{D^*} = \frac{7m}{2.144m} = 3.3$$

3. Compare H/D^* with the corresponding values listed in Table 2-5.

The value of 3.3 is not within the range of 3.6 – 16 listed in Table 2-5 for H/D^*.

Ceiling Jet Radial Distance: For scenarios involving a target in the ceiling jet, it is important to evaluate if the location of the target is within the range evaluated in this V&V study. The parameter r_{cj}/H can be used for such evaluation. In this normalized parameter, r_{cj} is the horizontal radial distance between the target and the centerline of the plume, and H is the vertical distance from the base of the fire to the target.

Example 3: Using the information in Example 1, let's assume that the cable target is in the ceiling jet, 2 m (6.6 ft) away from the centerline of the fire plume. Determine if the target location in the ceiling jet is within the locations included in this V&V report.

1. Determine the r_{cj} and H values from the information in Examples 1 & 3

   $r_{cj} = 2m \quad H = 7m$

2. Calculate r_{cj}/H

   $$\frac{r_{cj}}{H} = \frac{2m}{7m} = 0.29$$

3. Compare r_{cj}/H with the corresponding values listed in Table 2-5.

   The value of 0.29 is not within the range of 1.2 – 1.7 listed in Table 2-5 for r_{cj}/H.

Radiative Heat Flux: The normalized parameter $(D/r)^2$ is recommended for determining if the distance from a target to the flames is within the range of distances included in this V&V study. This parameter results from the dimensionless point source model $(X_r Q^*_d)/(4(r/D)^2)$. The Q^* term is removed since this is evaluated individually. This is important in scenarios involving targets affected by flame radiation. In the normalized parameter $(D/r)^2$, D is the fire diameter, and r is the horizontal distance from the targets to the flames.

Example 4: Using the information in Example 1, let's assume now that the cable target is in a vertical cable tray along a wall 2 m (6.6 ft) from the fire. The analyst is interested in investigating if flame radiation in the lower layer can affect the cable tray. Determine if the target location is within the locations included in this V&V report.

1. Determine the fire diameter from the information provided in Example 1

   $$D = \sqrt{\frac{4A_{dike}}{\pi}} = \sqrt{\frac{4(5.3m^2)}{3.1415}} = 2.5977 \approx 2.6m \quad r_{rad} = 2m$$

2. Calculate $(D/r_{rad})^2$

   $$\left(\frac{D}{r_{rad}}\right)^2 = \left(\frac{2.6m}{2m}\right)^2 = 1.7$$

3. Compare $(D/r_{rad})^2$ with the corresponding values listed in Table 2-5.

   The value of 1.7 not within the range of 0.03 – 0.2 listed in Table 2-5 for $\left(\frac{D}{r_{rad}}\right)^2$.

Room Size: Many fire scenarios require the specification of room size as an input for estimating fire-generated conditions inside an enclosure. The normalized parameters W/H and L/H can be used for determining if the room size in the scenario under evaluation is within the range of room sizes considered in this study. W and L are room width and length respectively. H is the height of the room.

Example 5: Continuing with Example 1, let's consider in this case that the cable tray is located in the hot gas layer. The hot gas layer temperature needs to be calculated for determining if the target may be damaged. The room is 8 m long and 6 m wide. Determine if the room size is within the room sizes included in this study.

1. Calculate W/H and L/H

$$\frac{W}{H} = \frac{6m}{6m} = 1 \qquad \frac{L}{H} = \frac{8m}{6m} = 1.3$$

2. Compare W/H and L/H with the corresponding values listed in Table 2-5.

   Values are within range listed in Table 2-5 for W/H and L/H.

Natural and Mechanical Ventilation: It is important to know whether a given compartment fire is limited by its fuel supply or by its oxygen supply. In all six test series, the fuel supply was specified as a test parameter. The oxygen supply was controlled either by the size of the compartment opening or by the flow rate of the ventilation system. Although less precise, it is possible to estimate the mass flow of oxygen for each test configuration. Where there is a door to the compartment, an estimate of the maximum achievable oxygen supply is given as follows:

$$\dot{m}_{O_2} = 0.23 \times \frac{1}{2} A_o \sqrt{h_o}$$

Where: $\dot{m}_{O_2}$ is the mass flow of oxygen (kg/s),

A_o is the area of the opening (m^2),

h_o is its height (m), and

0.23 is the mass fraction of oxygen in air.

Note that in many of the test series under consideration, this theoretical mass flow of oxygen was not achieved because the fires were of short duration. However, the estimate is useful for this exercise.

For an active ventilation system, the mass flow rate of oxygen is approximated by the following equation:

$$\dot{m}_{O_2} = 0.23 \times \rho_{air} \dot{V}$$

Where: ρ_{air} is the density of the air (kg/m^3), and

$\dot{V}$ is the ventilation rate (m^3/sec)

Of course, not all of the air supplied by the ventilation system would reach the fire, especially for ceiling-mounted supply ducts.

The global equivalence ratio, ϕ, is the ratio of the mass flow of fuel to the mass flow of oxygen, normalized by the stochiometric ratio. The estimated values of ϕ in the table below are based on the maximum fuel and oxygen flow rates for the given test series. The values are all less than one, meaning that these fires would be characterized as *well-* or *over-ventilated*. The test with the highest value of ϕ is ICFMP BE #4, Test 1, which has an equivalence ratio of about 0.6.

It is also notable in that its HRR to volume ratio is about an order of magnitude larger than all the other test series. Although still over-ventilated, the fire is relatively large with respect to its compartment volume. All of the other fires could be characterized as relatively small with respect to the compartment.

Example 6: Using the information in Examples 1 and 5, let's assume now that the cable target is in the hot gas layer, and the room is filled with smoke. The analyst is interested in investigating if the ventilation conditions and amount of oxygen available for combustion in the room are consistent with the conditions included in the V&V report. First, assume the room is naturally ventilated with a normally opened door. Then, assume the room is mechanically ventilated with 8 air changes per hour.

1. Determine ϕ, Natural Ventilation parameter, assuming an opening of 2m x 2m

$$\phi = \frac{\dot{Q}}{m_{o_2} \cdot (13100)} \qquad m_{o_2} = 0.23 \times \tfrac{1}{2} A_o \sqrt{h_o} = 0.23 \times \tfrac{1}{2}(4m^2)\sqrt{2m} = 0.650$$

$$\phi = \frac{7416kW}{(0.650)(13100)} = 0.870$$

2. Determine ϕ, Mechanical Ventilation parameter
   (room parameters; H=7m, W=6m, L=8m)

$$m_{o_2} = 0.23 \times \rho_{air}\dot{V} = 0.23(1.18kg/m^3)\left[(7m \times 6m \times 8m)(8ach)/3600\right] = 0.2026kg/s$$

$$\phi = \frac{\dot{Q}}{m_{o_2} \cdot (13100)} = \frac{7416kW}{(0.2026kg/s)(13100)} = 2.8$$

3. Compare ϕ, Natural Ventilation and ϕ, Mechanical Ventilation with the corresponding values listed in Table 2-5.
   For the mechanical ventilation, the value of 2.8 is outside the range listed in Table 2-5. In the natural ventilation case, the value is within the range listed Table 2.5.

Table 2-5: Summary of the Fire Experiments in Terms of Commonly Used Metrics

Parameter	ICFMP BE #2	ICFMP BE #3	ICFMP BE #4	ICFMP BE #5	FM/SNL	NBS Multi-Room	Validation Range
Q (kW)	1800 - 3600	400 - 2300	3500	400	500	100	N/A
$\dot{Q}_d^*$	≈ 1.0	0.4 - 2.1	2.4	0.7	0.6	1.4	0.4 – 2.4
ϕ, Natural Ventilation	0.6	0.14[#]	0.6	0.1	-	0.04	0.04 – 0.6
ϕ, Mechanical Ventilation	0.1	0.6	-	-	0.04 - 0.4	-	0.04 – 0.6
$D(m)$	1.17 - 1.60	1.00	1.13	0.79	0.91	0.34	N/A
$H(m)$	15.9	3.82	5.7	5.7	6.1	2.16	N/A
H/D	12 - 16	2.9 - 5.7	3.6	8.6	8.4	5.6	3.6 – 16
r_d	-	4.8 - 6.3	-	-	7.1 - 9.8		N/A

Parameter	ICFMP BE #2	ICFMP BE #3	ICFMP BE #4	ICFMP BE #5	FM/SNL	NBS Multi-Room	Validation Range
r_{ci}/H	-	1.3 - 1.7	-	-	1.2 - 1.6	-	1.2 – 1.7
W/H	0.7	5.7	0.6	0.6	3	1.1	0.6 – 5.7
L/H	1.4	1.8	0.6	0.6	2	1.1	0.6 - 2.0
$r_{rad}(m)$	-	2.2 - 5.7	-	-	-	-	N/A
$\left(D/r_{rad}\right)^2$	-	0.03-0.2	-	-	-	-	0.03 – 0.2

[#]Calculated for open door tests with no mechanical ventilation. An equivalence ratio was not calculated for closed door rooms since the tests were stopped when the oxygen concentration in the room reached 14%. That is, the fire was ventilated with the air in the test compartment during the duration of the experiment.

2.6.2 Characterization of Fire Model Predictions Based on Validation Results

Once the user determines the validation results reported here are applicable (see Section 2.6.1), the user must determine the predictive capability of the fire models. ASTM E 1355 does not provide specific criteria by which to judge the predictive capability of the models based on the results of the V&V. As such, the V&V project team developed a grading criteria and methodology to judge the models' capabilities. The criteria the team used are described below.

The process of deducing the predictive capability of the model from the quantitative results is documented in Volumes 3 through 7 of this report. Appendix A to each volume contains detailed comparisons of model prediction and experimental measurements. Chapter 6 of these volumes describes how these quantitative results were used to arrive at the characterization of predictive capability of the model using the approach described below.

The following two criteria are used to characterize the predictive capability of the model:

Criterion 1: *Are the physics of the model appropriate for the calculation being made?* This criterion reflects an evaluation of the underlying physics described by the model and the physics of the fire scenario. Generally the scope of this study is limited to the fire scenarios that are within the stated capability of the selected fire models (e.g., this study does not address the fire scenarios that involve flame spread within single and multiple cable trays).

Criterion 2: *Are there calculated relative differences outside the experimental and model input uncertainty?* This criterion is used as an indication of the accuracy of the model prediction. Since fire experiments are used as a way of establishing confidence in model prediction, the confidence can only be as good as our experiments and the model inputs derived from experiments. Therefore, if model predictions fall within the ranges of these combined uncertainties, the predictions are determined to be as accurate as the experiments and data.

Section 2.6.3 and Volume 2 of this report series provide an introduction and technical details for the uncertainty analysis.

The predictive capability of the model is characterized as follows based on the above criteria:

GREEN: If both criteria are satisfied (i.e., the model physics are appropriate for the calculation being made and the calculated relative differences are within or very near experimental uncertainty), then the V&V team concluded that the fire model prediction is accurate for the ranges of experiments in this study, and as described in Tables 2-4 and 2-5. A grade of GREEN indicates the model can be used with confidence to calculate the specific attribute. The user should recognize, however, that the accuracy of the model prediction is still somewhat uncertain and for some attributes, such as smoke concentration and room pressure, these uncertainties may be rather large. It is important to note that a grade of GREEN indicates validation only in the parameter space defined by the test series used in this study; that is, when the model is used within the ranges of the parameters defined by the experiments, it is validated.

YELLOW±: If the first criterion is satisfied and the calculated relative differences are outside the experimental uncertainty but indicate a consistent pattern of model over-prediction or under-prediction, then the model predictive capability is characterized as YELLOW+ for over-prediction, and YELLOW– for under-prediction. The model prediction for the specific attribute may be useful within the ranges of experiments in this study, and as described in Tables 2-4 and 2-5, but the users should use caution when interpreting the results of the model. A complete understanding of model assumptions and scenario applicability to these V&V results is necessary. The model may be used if the grade is YELLOW+ when the user ensures that model over-prediction reflects conservatism. The user must exercise caution when using models with capabilities described as YELLOW±.

YELLOW: If the first criterion is satisfied and the calculated relative differences are outside experimental uncertainty with no consistent pattern of over- or under-prediction, then the model predictive capability is characterized as YELLOW. A YELLOW classification is also used despite a consistent pattern of under- or over-prediction if the experimental data set is limited. Caution should be exercised when using a fire model for predicting these attributes. In this case, the user is referred to the details related to the experimental conditions and validation results documented in Volumes 2 through 6. The user is advised to review and understand the model assumptions and inputs, as well as the conditions and results to determine and justify the appropriateness of the model prediction to the fire scenario for which it is being used.

RED: If the first criterion is not met, then the particular fire model capability should not be used.

No color: This V&V study did not investigate this capability. This may be attributable to one or more reasons that include unavailability of appropriate data or lack of model, sub-model, or output.

As suggested in the criteria above, there is a level of engineering judgment in the classification of fire model predictive capabilities. Specifically, the V&V project team exercised engineering judgment in the following two areas:

1. Evaluation of the modeling capabilities of the particular tool if the model physics are appropriate.

2. Evaluation of the magnitude of relative differences when compared to the experimental uncertainty. Judgment in this area impacts the determination of Green versus Yellow color.

The team included fire model developers, NPP fire modeling experts, and code users. In general, a Green or Yellow classification suggests that the V&V team determined that the model physics are appropriate for the calculation been made, within the assumptions of the specific model. The difference between the colors is attributable to the magnitude of the calculated relative differences. Judgment considerations include general experimental conditions, experimental data quality, and the characterization of the experimental uncertainty.

2.6.3 Uncertainty Analysis

The relative differences between experimental observations and model predictions are compared in terms of a pre-determined combined uncertainty on which the validation results for each fire modeling attribute are determined. This combined uncertainty includes the contribution of two sources of uncertainty:

- Experimental uncertainty, which is the uncertainty associated with the measurement devices in the different experiments.

- Model input uncertainty, which refers to the uncertainty in the inputs to the model. Notice that the inputs to the fire models, such as the heat release rate, are (for the most part) measured parameters in the different experiments.

Consider as an example, comparisons between measured and predicted values in terms of the range of the combined uncertainty (scatter plot in Figure 2-10a). If the models and experiments were in perfect agreement, the data points would fall along the 45° line indicated in the figure. The data in Figure 2-10a are such that the models are in agreement with the experiments within the bounds of the combined uncertainties. In contrast, Figure 2-10b shows data in which the models and experiments are not in agreement within the bounds of the combined uncertainties. In this case, the comparison suggests that some aspect of the uncertainty was not captured by the combined uncertainty value.

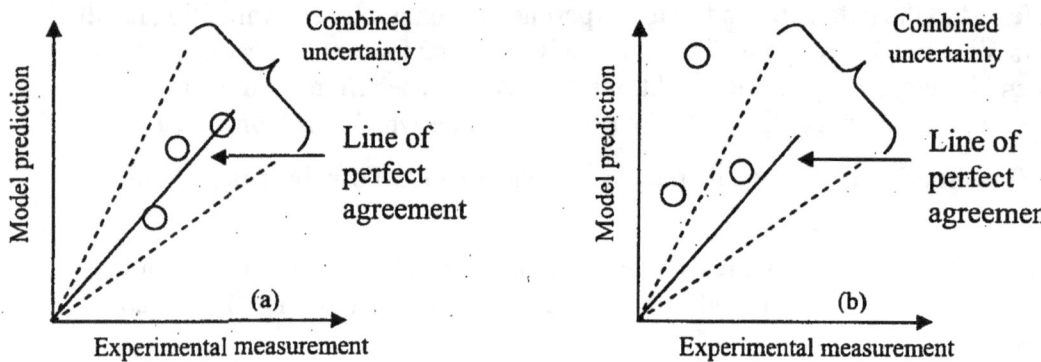

Figure 2-10: Scatter plots depicting validation results in and out of the range of combined uncertainty.

Volume 2 describes in detail the quantification of the combined uncertainty for each fire modeling attribute.

3
RESULTS AND CONCLUSIONS

This chapter summarizes the results and findings of this V&V study. Section 3.1 presents the results using the approach discussed in Section 2.6.2 of this volume. Readers should also review Chapter 6 of Volumes 3 through 7 for more detailed discussions of the quantitative results. In addition, Appendix A to each volume presents a graphical representation of the experimental observations and outputs from the models.

Section 3.2 discusses observations concerning the results. These observations highlight patterns and relationships between the results presented and experiments. Section 3.3 contains our conclusion regarding our findings.

3.1 Results

3.1.1 Validation

Table 3-1 summarizes the results of this validation study. As discussed in Section 2.6.2, the predictive capabilities of the models are graded based on the quantitative values of relative difference between model prediction and experimental measurements.

The validation results show that none of the models have attributes that are RED. This is because all the models appropriately represent the physics of the scenarios, within the simplifying assumptions of the calculation method. Most of the correlations employed within the models were empirically confirmed theoretical derivations of general physical phenomena, as summarized in Chapter 3 of each volume. All of the relative differences that fell significantly outside of the combined uncertainty could be explained in terms of the simplifying assumptions within the models and the comparison of these assumptions with the experimental configurations.

Table 3-1: Results of the Validation & Verification of the Selected Fire Models for Nuclear Power Plant Fire Modeling Applications

Parameter[5]		FDT[S]	FIVE-Rev1	CFAST	MAGIC	FDS
Hot gas layer temperature ("upper layer temperature")	Room of Origin	YELLOW-	YELLOW+	GREEN	GREEN	GREEN
	Adjacent Room	N/A	N/A	YELLOW	YELLOW+	GREEN
Hot gas layer height ("layer interface height")		N/A	N/A	GREEN	GREEN	GREEN
Ceiling jet temperature ("target/gas temperature")		N/A	YELLOW+[2]	YELLOW+	GREEN	GREEN
Plume temperature		YELLOW-	YELLOW+[2]	N/A	GREEN	YELLOW
Flame height[3]		GREEN	GREEN	GREEN	GREEN	YELLOW[1]
Oxygen concentration		N/A	N/A	GREEN	YELLOW	GREEN
Smoke concentration		N/A	N/A	YELLOW	YELLOW	YELLOW
Room pressure[4]		N/A	N/A	GREEN	GREEN	GREEN
Target temperature		N/A	N/A	YELLOW	YELLOW	YELLOW
Radiant heat flux		YELLOW	YELLOW	YELLOW	YELLOW	YELLOW
Total heat flux		N/A	N/A	YELLOW	YELLOW	YELLOW
Wall temperature		N/A	N/A	YELLOW	YELLOW	YELLOW

Parameter[5]	Fire Model				
	FDT[s]	FIVE-Rev1	CFAST	MAGIC	FDS
Total heat flux to walls	N/A	N/A	YELLOW	YELLOW	YELLOW

Notes:

1. FDS does not use an empirical correlation to predict the flame height. Rather, it solves a set of equations appropriate for reacting flows and predicts the flame height as the uppermost extent of the combustion zone. This is a challenging calculation and the Yellow emphasizes that caution should be exercised by users.

2. FIVE approximates the experimental plume as the sum of hot gas layer temperature and the calculated plume temperature and experimental ceiling jet temperature as the sum of hot gas layer temperature and the calculated ceiling jet temperature. The calculated plume and ceiling jet temperatures were obtained from the correlations.

3. Flame height models compared with visual observations only.

4. Large experimental uncertainties for room pressure.

5. Refer to Table 2-3 for information about which experiments captured data for each parameter.

3.1.2 *Verification*

All five models have been verified by this study as appropriate for fire protection applications, within the assumptions for each individual model or sub-model. The project team used guidance in ASTM E 1355 about the theoretical basis and mathematical and numerical robustness to make this determination. The verification for each model is documented in Volumes 3 through 7.

During the process of this study, a number of modifications and corrections to the five selected fire models were identified and implemented. These modification and corrections were identified during the validation as a result of trying to interpret the results. The nature of these modifications and corrections cover a wide range from inconsequential to those that could lead to incorrect result. Descriptions of these modifications can be found in Volume 3 through 7.

3.2 Summary

This section presents a summary of observations from the results of this study. These observations are based on review of these results and generally apply to the five fire models considered in this study:

- The experiments considered in this study represent configurations that may be seen in NPP applications. Not all possible NPP scenarios are evaluated in this study. We have evaluated only those scenarios in which suitable experimental data is available. For a variety of reasons, however, the experimental data is limited. Users should evaluate independently whether the results of this study are applicable to their specific scenario. Table 2-5 provides guidance in this matter.

- For the fire scenarios considered in the current validation study, and for the output quantities of interest, the libraries of engineering calculations (FDT^s, FIVE-Rev1) have limited capabilities. These libraries do not have appropriate methods for estimating many of the fire scenario attributes evaluated in this study. The correlations that the libraries do contain are typically empirically deduced from a broad database of experiments. The correlations are based on fundamental conservation laws and have gained a considerable degree of acceptance in the fire protection engineering community. However, because of their empirical nature, they are subject to many limiting assumptions. The user must be cautious when using these tools.

- The two-zone models performed well when compared with the experiments considered. Evaluation of the two-zone models showed that the models simulated the experimental results within experimental uncertainty (see Chapter 6 in Volume 2) for most of the parameters of interest The reason for this may be that the relatively simple experimental configurations selected for this study conform well to the simple two-layer assumption that is the basis of these models. However, users must remain cautious when applying these models to more complex scenarios, or when predicting certain phenomena, like heat fluxes.

- Evaluation of the FDS model showed that the model simulated the experimental results within experimental uncertainty (see Chapter 6 in Volume 2) for most of the parameters of interest. The results of the field model, FDS, are comparable to the results of the two-zone models (CFAST, Volume 5, and MAGIC, Volume 6), probably because the experimental configurations utilized in this study contained, in most cases, two distinct layers within the room of origin.

- The decision to use any of these models can depend on many considerations. Real fire scenarios rarely conform neatly to some of the simplifying assumptions inherent in the models. Although engineering calculations and two-zone models can be applied in instances where the physical configuration is complex, their accuracy cannot be ensured. Field model predictions can be more accurate in more these complex scenarios. However, the time it takes to get and understand a prediction may also be an important consideration in the decision to use a particular model for a specific scenario. FDS is computationally expensive and, while the two-zone models produce answers in seconds to minutes, FDS provides comparable answers in hours to days. FDS is better suited to predict fire environments within more complex configurations because it predicts the local effects of a fire.

- Like all predictive models, the best predictions come with a clear understanding of the limitations of the model and of the inputs provided to do the calculations. For calculation of many attributes (see those attributes categorized as yellow in Table 3-2), caution should be exercised when applying these models. For the attributes categorized as GREEN, the models are accurate to within the experimental uncertainty associated with each particular attribute (see Chapter 6 in Volume 2) for the range of conditions (see Table 2-5) represented by the experiments used in this study.

3.3 Conclusions

This study provides justification for verification and provides validation via comparisons between experimental data and predictions from five fire modeling tools. The validation results of this study are presented in the form of color-coded grades of the predictive capability of fire models for important parameters for NPP fire modeling applications. These grades are based on the quantitative relative differences between model predictions and applicable experimental measurements. The predictive capability considers the uncertainty in the experimental measurements.

The use of fire models to support fire protection decision-making requires understanding of their limitations and confidence in their predictive capabilities. This report improves the understanding and evaluates the predictive capabilities of the models selected. Fully understanding the predictive capabilities of fire models is a challenge that should be addressed if the fire protection community is to realize the full benefit of fire modeling. The approach used in this study and documented and implemented in the individual volumes can be used as a roadmap to model users and developers for conducting a V&V for models other than those included in this study.

The results of this project clearly suggest that any fire modeling analysis should consider the predictive capabilities associated with the analytical tool when interpreting its results.

4
REFERENCES

1. *Performance-Based Standard for Fire Protection for Light-Water Reactor Electric Generating Plants*, NFPA 805, 2001 Edition, National Fire Protection Association, Brainerd, MA, 2001.

2. "Voluntary Fire Protection Requirements for Light-Water Reactors," 10 CFR Part 50, Section 50.48(c), RIN 3150-AG48, *Federal Register*, Volume 69, Number 115, U.S. Nuclear Regulatory Commission, Washington, DC, June 16, 2004.

3. Memorandum of Understanding (MOU) on Cooperative Nuclear Safety Research Between EPRI and NRC, Amendment on Fire Risk, U.S. Nuclear Regulatory Commission, Washington, DC, Revision 1, 2001.

4. *Standard Guide for Evaluating Predictive Capability of Deterministic Fire Models,* ASTM E1355-05a, American Society for Testing and Materials, West Conshohocken, PA, 2004.

5. *Fire Modeling Guide for Nuclear Power Plant Applications,* EPRI 1002981, Electric Power Research Institute, Palo Alto, CA, August 2002.

6. EPRI 1011989 and NUREG/CR-6850, "EPRI/NRC-RES Fire PRA Methodology for Nuclear Power Facilities," U.S. Nuclear Regulatory Commission, Washington, DC, August 2005.

7. NUREG-1805, "Fire Dynamics Tools (FDT^s): Quantitative Fire Hazard Analysis Methods for the U.S. Nuclear Regulatory Commission Fire Protection Inspection Program," U.S. Nuclear Regulatory Commission, Washington, DC, December 2004.

8. Jones, W., R. Peacock, G. Forney, and P. Reneke, "CFAST: An Engineering Tool for Estimating Fire and Smoke Transport, Version 5 — Technical Reference Guide," SP 1030, National Institute of Standards and Technology, Gaithersburg, MD, 2004.

9. Gay, L., and C. Epiard, "MAGIC Software Version 4.1.1: Mathematical Model" EDF HI82/04/024/P, December 2004.

10. "Fire Dynamics Simulator (Version 4) Technical Reference Guide," K. McGrattan, ed., National Institute of Standards and Technology, Special Publication 1018, 2004.

11. NUREG/CR-4681, "Enclosure Environment Characterization Testing for the Base Line Validation of Computer Fire Simulation Codes," U.S. Nuclear Regulatory Commission, Washington, DC, March 1987.

12. NUREG/CR-5384, "A Summary of Nuclear Power Plant Fire Safety Research at Sandia National Laboratories, 1975–1987," U.S. Nuclear Regulatory Commission, Washington, DC, December 1989.

13. NBSIR 88-3752, " An Experimental Data Set for the Accuracy Assessment of Room Fire Models," National Bureau of Standards, April 1988

14. EPRI 108875, "Fire Modeling Code Comparison," Electric Power Research Institute, Palo Alto, CA, September 1998.

15. Hostikka, S., M. Kokkala, and J. Vaari, "Experimental Study of the Localized Room Fires," NFDC2 Test Series, VTT Research Notes 2104, 2001.

16. Hamins, A., A. Maranghides, E. Johnsson, M. Donnelly, J. Yang, G. Mulholland, and R. Anleitner, *Report of Experimental Results for the International Fire Model Benchmarking and Validation Exercise #3*, NIST Special Publication 1013-1, National Institute of Standards and Technology, Gaithersburg, MD, 2005.

17. Klein-Heßling, W., and M. Röwenkamp, *Evaluation of Fire Models for Nuclear Power Plant Applications: Fuel Pool Fire inside a Compartment*, Gesellschaft für Anlagen-und Reaktorsicherheit (GRS), Köln, Germany, May 2005.

18. Hosser, D., O. Riese, and M. Klingenberg, *Performing of Recent Real Scale Cable Fire Experiments and Presentation of the Results in the Frame of the International Collaborative Fire Modeling Project ICFMP*, Institut für Baustoffe, Massivbau und Brandschutz (iBMB), Braunschweig, Germany, June 2004.

19. Karlsson, B., and J.G. Quintiere, "Enclosure Fire Dynamics," CRC Press, Boca Raton, FL, 2000.

20. Zukoski E.E., "Properties of Fire Plumes," *Combustion Fundamentals of Fire*, G. Cox, ed., Academic Press, Burlington, MA, 1995.

NRC FORM 335
(9-2004)
NRCMD 3.7

U.S. NUCLEAR REGULATORY COMMISSION

BIBLIOGRAPHIC DATA SHEET

(See instructions on the reverse)

1. REPORT NUMBER
(Assigned by NRC, Add Vol., Supp., Rev., and Addendum Numbers, if any.)

NUREG-1824

2. TITLE AND SUBTITLE

Verification and Validation of Selected Fire Models for Nuclear Power Plant Applications
Volume 1: Main Report

3. DATE REPORT PUBLISHED

MONTH	YEAR
May	2007

4. FIN OR GRANT NUMBER

5. AUTHOR(S)

B. Najafi (EPRI/SAIC), F. Joglar (EPRI/SAIC), J. Dreisbach (NRC)

6. TYPE OF REPORT

Technical

7. PERIOD COVERED *(Inclusive Dates)*

8. PERFORMING ORGANIZATION - NAME AND ADDRESS *(If NRC, provide Division, Office or Region, U.S. Nuclear Regulatory Commission, and mailing address; if contractor, provide name and mailing address.)*

U.S. Nuclear Regulatory Commission, Office of Regulatory Research (RES), Washington, DC 20555-0001

Electric Power Research Institute (EPRI), 3412 Hillview Avenue, Palo Alto, CA 94303

Science Applications International Corp. (SAIC), 4920 El Camino Real, Los Altos, CA 94022

National Institute of Standards and Technology (NIST/BFRL), 100 Bureau Drive, Stop 8600, Gaithersburg, MD 20899-8600

9. SPONSORING ORGANIZATION - NAME AND ADDRESS *(If NRC, type "Same as above"; if contractor, provide NRC Division, Office or Region, U.S. Nuclear Regulatory Commission, and mailing address.)*

U.S. Nuclear Regulatory Commission, Office of Regulatory Research (RES), Washington, DC 20555-0001

Electric Power Research Institute (EPRI), 3412 Hillview Avenue, Palo Alto, CA 94303

10. SUPPLEMENTARY NOTES

11. ABSTRACT *(200 words or less)*

There is a movement to introduce risk-informed and performance-base analyses into fire protection engineering practice, both domestically and worldwide. The move towards risk-informed decision-making in nuclear power regulation was directed by the U.S. Nuclear Regulatory Commission.

One key tool needed to support risk-informed, performance-based fire protection is the availability of verified and validated fire models that can accurately predict the consequences of fires. Section 2.4.1.2. of NFPA 805, Performance-Base Standard for Fire Protection for Light-Water Reactor Electric Generating Plants, 2001 Edition requires that only fire models acceptable to the Authority Having Jurisdiction (AHJ) shall be used in fire modeling calculations. Futhermore, Sections 2.4.1.2.2. and 2.4.1.2.3. of NFPA 805 state that fire models shall be applied within the limitations of the given model, and shall be verified and validated. This report is the first effort to document the verification and validation (V&V) of five models that are commomly used in nuclear power plant applications. The project was performed in accordance with the guidelines that the American Society for Testing and Materials (ASTM) set forth in ASTM E 1355, Standard Guide for Evaluating the Predictive capability of Deterministic Fire Models. The results of this V&V are reported in the form of color codes describing the accuracies for the model predictions.

12. KEY WORDS/DESCRIPTORS *(List words or phrases that will assist researchers in locating the report.)*

fire, fire modeling, verification, validation, performance-based, risk-informed, firehazards analyses, V&V, FHA, CFAST, FDS, MAGIC, FIVE, FDTs

13. AVAILABILITY STATEMENT

unlimited

14. SECURITY CLASSIFICATION

(This Page)

unclassified

(This Report)

unclassified

15. NUMBER OF PAGES

16. PRICE

www.ingramcontent.com/pod-product-compliance
Lightning Source LLC
Chambersburg PA
CBHW081606170526
45166CB00009B/2850